Design of Phase-Locked Loop Circuits, With Experiments

by
Howard M. Berlin

Also Published as
The Phase-Locked Loop Bugbook,® With Experiments
by E & L Instruments, Inc.

fff
HOWARD W. SAMS & COMPANY
A Division of Macmillan, Inc.
4300 West 62nd Street
Indianapolis, Indiana 46268 USA

© 1978 by Howard M. Berlin

FIRST EDITION
TWELFTH PRINTING — 1989

All rights reserved. No part of this book shall be reproduced, stored in a retrieval system, or transmitted by any means, electronic, mechanical, photocopying, recording, or otherwise, without written permission from the publisher. No patent liability is assumed with respect to the use of the information contained herein. While every precaution has been taken in the preparation of this book, the publisher assumes no responsibility for errors or omissions. Neither is any liability assumed for damages resulting from the use of the information contained herein.

International Standard Book Number: 0-672-21545-3
Library of Congress Catalog Card Number: 78-57203

Printed in the United States of America.

Preface

With the rapid evolution of integrated-circuit technology, the phase-locked loop has established itself as one of the fundamental building blocks in the electronics revolution. In the early 1970s, when the phase-locked loop was built using transistors, the cost of the circuit alone was enough to discourage widespread applications. Only now is it possible to fully appreciate what the phase-locked loop can do.

Although there are a number of textbooks covering the details of the phase-locked loop, they are for the most part highly mathematical. To my knowledge, there is presently no book covering the principles of the phase-locked loop based on integrated circuits in conjunction with a wide range of laboratory-type experiments. This book was written in an effort to fill this void.

Using both TTL and CMOS integrated circuits, this text/workbook covers the operation of the phase detector, voltage-controlled oscillator, loop filter, frequency synthesizers, and monolithic systems, with applications, in seven chapters. In addition, there are over 15 experiments that demonstrate the concepts presented throughout the book. For this reason, the book is useful to the experimenter and hobbyist who wants to learn by self-study, or it can easily serve as an addition to any of the college courses on control systems or linear integrated circuits, especially those which have a laboratory section. A strong attempt has been made to keep the use of mathematical equations to a bare minimum, giving only the essential relationships. Any calculations can easily be performed using a simple pocket calculator. However, the derivation of the major equations and design criteria are presented in Appendix A, which I hope will satisfy some of you.

This is my fourth book in the *Blacksburg Continuing Education Series*™. The other three are: *555 Timer Applications Sourcebook, With Experiments; Design of Active Filters, With Experiments;* and *Design of Op-Amp Circuits, With Experiments.*

Finally, there are a number of individuals and manufacturers without whose assistance this book would not be possible. I would like to thank David Larsen and Peter Rony of the Virginia Polytechnic Institute and State University and Jon and Chris Titus of Tychon, Inc. for their valuable advice and assistance. Gratitude is also extended to E&L Instruments, Inc., who continue to support my efforts, and to Hughes Aircraft Co. (Solid State Products Division), Motorola Semiconductor Products, Inc., RCA Solid State Division, and Signetics Corporation for allowing me to reproduce technical data from their promotional literature and catalogs.

Howard M. Berlin, W3HB

Howard M. Berlin is an electrical engineer with the Chemical Systems Laboratory at Aberdeen Proving Ground, Maryland, and has been an adjunct instructor in the Department of Electrical Engineering at the University of Delaware. His experience has primarily been in biomedical engineering research and physiological instrumentation. He has taught several short courses for the Department of the Army, several universities and conferences, and graduate courses at the University of Delaware. He has authored a number of governmental reports and articles in scientific and amateur radio magazines. In addition, he is the author of the following books: **555 Timer Applications Sourcebook, With Experiments; Design of Active Filters, With Experiments; Design of Op-Amp Circuits, With Experiments; Guide to CMOS Basics, With Experiments;** and co-author of **Design of VMOS Circuits, With Experiments.** He is presently a member of Sigma XI, IEEE, and the Delaware Academy of Medicine. As an active ham radio operator, he can be heard using the call letters W3HB, and was formerly K3NEZ.

This book is dedicated to the memory of my father.

Contents

CHAPTER 1

THE BASIC PHASE-LOCKED LOOP PRINCIPLE 7
Introduction—Objectives—The Basic Principle—A Brief History of the Phase-Locked Loop

CHAPTER 2

PERFORMING THE EXPERIMENTS 11
Introduction—Rules for Setting Up Experiments—Format for the Experiments—How Many Experiments Do I Perform?—Breadboarding—Helpful Hints and Suggestions—Equipment—Input/Output Circuits—Components

CHAPTER 3

THE PHASE DETECTOR 24
Introduction—Objectives—Phase—The Phase Detector—The Exclusive-OR Phase Detector—Edge-Triggered Phase Detectors—The MC4044 Phase Detector—An Introduction to the Experiments—Experiments

CHAPTER 4

THE VOLTAGE-CONTROLLED OSCILLATOR 53
Introduction—Objectives—VCO Basics—VCO Circuits—The Varactor—Other Integrated Circuits—An Introduction to the Experiment—Experiment

CHAPTER 5

The Loop Filter and Loop Response 63

Introduction—Objectives—Function of the Loop Filter—Low-Pass Filter Circuits—The Transient Response—Lock and Capture—An Introduction to the Experiment—Experiment

CHAPTER 6

Digital Frequency Synthesizers 78

Introduction—Objectives—The Basic Synthesizer—Practical Synthesizers—The Synthesizer Loop Filter—Frequency Reference Circuits—Divide-by-N Counters — TTL Fixed Counters — CMOS Fixed Counters—TTL Programmable Counters—CMOS Programmable Counters—Programming Switches—An Introduction to the Experiments—Experiments

CHAPTER 7

Monolithic Integrated Circuits and Applications . . 122

Introduction—Objectives—The 560 Series—The 4046 CMOS Phase-Locked Loop—An Introduction to the Experiments—Experiments

APPENDIX A

Derivations 151

APPENDIX B

Data Sheets 158

APPENDIX C

Breadboarding Aids 243

APPENDIX D

Symbols Used 249

Bibliography 250

Index 253

CHAPTER 1

The Basic Phase-Locked Loop Principle

INTRODUCTION

This chapter is a brief introduction to the phase-locked loop, which will acquaint you briefly with the loop's building blocks: the phase detector, loop filter, and voltage-controlled oscillator. Each of these components, however, will be discussed in greater detail in subsequent chapters.

OBJECTIVES

At the completion of this chapter, you will be able to do the following:

- Draw a block diagram of the basic phase-locked loop.
- Explain the general principle of the phase-locked loop.
- Briefly explain the function of the following:
 phase detector
 loop filter
 voltage-controlled oscillator

THE BASIC PRINCIPLE

As illustrated in the block diagram of Fig. 1-1, the phase-locked loop is basically an electronic feedback loop system consisting of:

1. A phase detector, or comparator.
2. A low-pass filter.
3. A voltage-controlled oscillator (vco).

From the study of feedback and control systems, these three components are said to be in the *forward path* of the loop, while the single connection between the vco and the phase detector is the *feedback path*.

The vco is a free-running oscillator, the frequency of which is normally determined by an external resistor-capacitor or an inductor-capacitor network. The vco frequency (f_o) is fed back to the phase detector where it is compared with the frequency of the input signal (f_i). The output of the phase detector is the *error voltage,* which is an average dc voltage proportional to the difference in frequency ($f_i - f_o$) and phase $\Delta\phi$ of the input and vco.

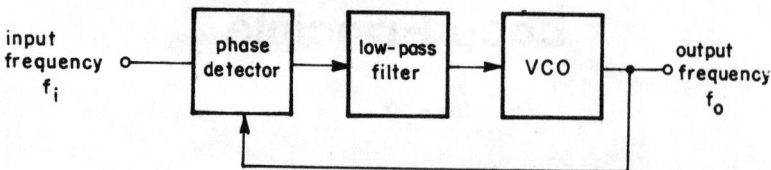

Fig. 1-1. Block diagram of the basic phase-locked loop.

The error voltage is then filtered, thus removing traces of higher frequency noise. This, in turn, is then fed to the vco to complete the loop. In addition, the error voltage forces the frequency of the vco to change in a direction that reduces the frequency difference between the input and the vco. Once the vco starts to change frequency, the loop is in the *capture state*. This process continues until the vco and the input frequencies are exactly the same. At this point, the loop is synchronized, or *phase-locked*. During phase-lock, the vco frequency is identical to the input of the loop, *except for a finite phase difference,* which is required to generate the necessary error voltage that shifts the vco frequency, keeping the loop in phase-lock. This repetitive action of the loop system then tracks, or follows, any change in the input frequency while phase-locked. We can say that the phase-locked loop has three distinct states:

1. Free-running.
2. Capture.
3. Phase-lock.

The range over which the loop system will follow changes in the input frequency is called the *lock range*. On the other hand, the frequency range in which the loop acquires phase-lock is the *capture range, and is never greater than the lock range.*

The dynamic characteristics of the phase-locked loop are controlled primarily by the low-pass filter. If the difference between the input and vco frequencies is significantly large, the resultant signal may be too high to be passed by the filter. Consequently, the signal is out of the capture range of the loop. Once the loop is phase-locked, the filter only limits the speed of the loop's ability to track changes in the input frequency. In addition, the loop filter provides a sort of short-term memory, ensuring a rapid recapture of the signal if the system is thrown out of lock by a noise transient.

In Chapter 5 we will see that the design of the loop filter represents a compromise. Although the parameters of the filter restrict the loop capture range and speed, it would almost be impossible for the phase-locked loop to lock without it.

A BRIEF HISTORY OF THE PHASE-LOCKED LOOP

In the early 1930s, the superheterodyne receiver was king. However, because of the number of tuned stages, a simpler method was desired. In 1932, a team of British scientists experimented with a method to surpass the superheterodyne. This new type of receiver, called the *homodyne* and later renamed the *synchrodyne,* first consisted of a local oscillator, a mixer, and an audio amplifier. When the input signal and the local oscillator were mixed at the same phase and frequency, the output was an exact audio representation of the modulated carrier. Initial tests were encouraging, but the synchronous reception after a period of time became difficult due to the slight drift in frequency of the local oscillator. To counteract this frequency drift, the frequency of the local oscillator was compared with the input by a phase detector so that a correction voltage would be generated and fed back to the local oscillator, thus keeping it on frequency. This technique had worked for electronic servo systems, so why wouldn't it work with oscillators? This type of feedback circuit began the evolution of the phase-locked loop. Although the synchronous, or homodyne, receiver was superior to the superheterodyne method, the cost of a phase-locked-loop circuit outweighed its advantages.

In the 1940s, the first widespread use of the phase-locked loop was in the synchronization of the horizontal and vertical sweep oscillators in television receivers to the transmitted sync pulses. Such circuits carried the names "*Synchro-Lock*" and "*Synchro-Guide.*" Since that time, the electronic phase-locked-loop principle has been extended to other applications. For example, radio telemetry data from satellites used narrow-band, phase-locked-loop receivers to recover low-level signals in the presence of noise. Other applications now include: am and fm demodulators, fsk decoders, motor speed controls, Touch-Tone® decoders, light-coupled analog isolators, and frequency synthesized transmitters and receivers. Several of these applications will be discussed in Chapters 6 and 7.

CHAPTER 2

Performing the Experiments

INTRODUCTION

Following this chapter, you will have the opportunity to perform a wide range of experiments covering all phases of the phase-locked loop using TTL and CMOS integrated-circuit devices. This chapter discusses the necessary equipment and components that you will need in order to perform the experiments easily and accurately.

RULES FOR SETTING UP EXPERIMENTS

Throughout this book, you will be breadboarding various circuits, either using some of the breadboarding aids manufactured by E&L Instruments, or constructing some of the necessary equipment. If you have already had experience with the *Blacksburg Continuing Education Series*™ texts, these rules will be familiar. Before you set any experiment, it is recommended that you do the following:

1. Plan your experiment beforehand. Know what types of results you are expected to observe.
2. Disconnect or turn off *all* power and external sources from the breadboard.
3. Clear the breadboard of all wires and components from previous experiments unless instructed otherwise.
4. Check the wired-up circuit against the schematic diagram to make sure that it is correct.
5. Connect or turn on the power and external signal sources to the breadboard *last!*

11

6. When finished, make sure that you disconnect everything *before* you clear the breadboard of wires and components.

FORMAT FOR THE EXPERIMENTS

The instructions for each experiment are presented in the following format:

Purpose

The material presented under this heading states the purpose of performing the experiment. It is well for you to have this intended purpose in mind as you conduct the experiment.

Pin Configuration of Integrated-Circuit Chips

The pin configurations are given under this heading for all of the integrated-circuit chips used in the experiment.

Schematic Diagram of Circuit

Under this heading is the schematic diagram of the completed circuit that you will wire up in the experiment. You should analyze this diagram in an effort to obtain an understanding of the circuit *before* you proceed further.

Design Basics

Under this heading is a summary of the design equations and/or characteristics that apply for the design and operation of the circuit.

Steps

A series of sequential steps describe the detailed instructions for performing portions of the experiments. Questions are also included at appropriate points. Any numerical calculations are performed easiest on many of the pocket-type calculators.

HOW MANY EXPERIMENTS DO I PERFORM?

In this text, there are many experiments. In several cases, a number of experiments are essentially repeated, differing in the type of integrated circuit used, such as TTL versus CMOS. Consequently, it is not necessary to perform every experiment. Some of you may want to experiment only with TTL integrated circuits rather than the more expensive CMOS devices. In either case, there are enough experiments for you to gain a good feeling for the operation of the phase-locked loop.

BREADBOARDING

The breadboard is designed to accommodate the many experiments that you will perform in the chapters to follow. The various integrated-circuit devices, resistors, capacitors, and other components, as well as electrical power, all connect or tie directly to the breadboard.

Shown in Fig. 2-1 is the top view of the basic component of a typical breadboarding system, which is known as the *SK-10 Universal Breadboarding Socket,* manufactured by E&L Instruments, Inc. It contains 64 by 2 sets of 5 electrically connected solderless terminals that straddle both sides of a narrow groove, and 8 sets of 25 electrically connected terminals along the edges. The center group of 5 electrically connected terminals accommodate the integrated-circuit chips and permit 4 additional connections to be made at each pin of the integrated circuit.

An extensive line of useful breadboarding aids manufactured by E&L Instruments is presented in Appendix C.

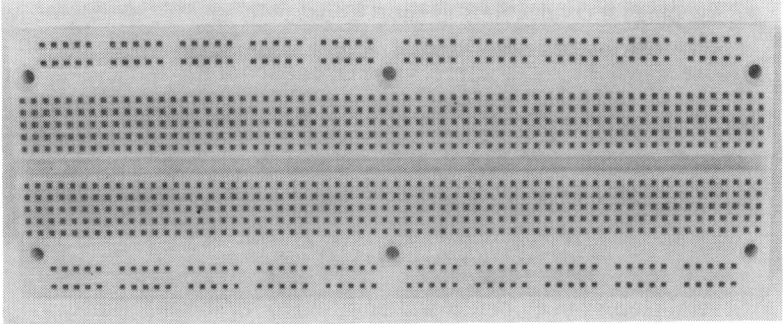

Courtesy E & L Instruments, Inc.

Fig. 2-1. SK-10 Universal Breadboarding Socket.

HELPFUL HINTS AND SUGGESTIONS

Tools

Only three tools are really necessary for all of the experiments given in this book:

1. A pair of "long-nosed" pliers.
2. A wire stripper/cutter.
3. A small screwdriver.

The pliers are used to:

- Straighten out the bent ends of hookup wire that is used to wire the circuits on the breadboard.

13

- Straighten out or bend the resistor, capacitor, and other component leads to the proper position so that they can be conveniently inserted into the breadboard.

The wire stripper/cutter is used to cut the hookup wire to size and strip about ⅜ inch of insulation from each end.

The screwdriver, if for nothing else, can be used to easily remove the integrated-circuit devices from the solderless breadboarding socket.

Wire

Only No. 22, No. 24, or No. 26 insulated wire is used, and it must be solid, not stranded!

Breadboarding

- Never insert too large a wire or component lead into a breadboard terminal.
- Never insert a bent wire. Straighten out the bent end with a pair of pliers before insertion.
- Try to maintain an orderly arrangement of components and wires, keeping all connections as short as possible.
- Plan the construction of your circuit on a layout sheet, like the one shown in Fig. 2-2, before you breadboard it.

Pocket Calculator

This is not mandatory, but it is recommended that you use one. The routine calculations can be accomplished quickly and accurately.

EQUIPMENT

Several pieces of equipment will be required for the experiments.

Oscilloscope

Just about any general-purpose type will do, but it must be at least a dual-trace type.

Frequency Counter

This doesn't have to be an expensive one, but it should have a resolution of 1 Hz for precise measurements. There are several low-cost units available in kit form for less than $80 which will perform nicely. Every serious experimenter should have one.

VOM, VTVM, or Digital Voltmeter

A general-purpose meter that is capable of measuring dc voltages is necessary. If you can obtain one, please use a digital type, as the resolution of the measurements will be better. If a vom is used, its

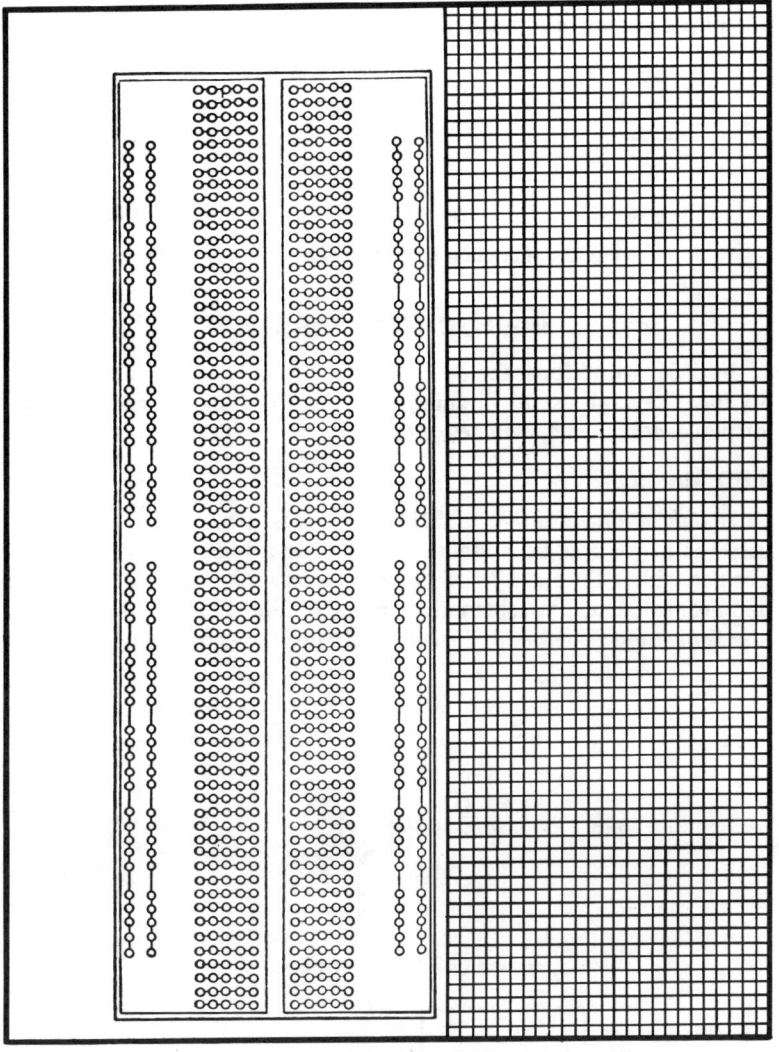

Courtesy E & L Instruments, Inc.
Fig. 2-2. SK-10 sketch sheet.

rating should be at least 20 kΩ/V so as not to introduce loading errors.

Function Generator

A function generator is capable of producing sine waves, square waves, and triangle waves with adjustable frequency and amplitude. Fig. 2-3 shows the schematic symbols that will be used to indicate the particular waveform used in the experiment.

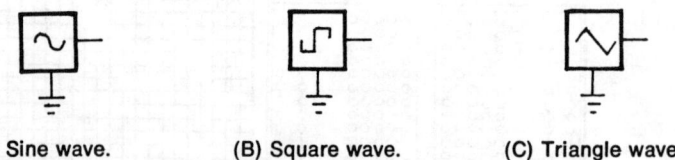

(A) Sine wave.   (B) Square wave.   (C) Triangle wave.

Fig. 2-3. Function generator schematic symbols.

INPUT/OUTPUT CIRCUITS

In order to perform the experiments smoothly, it will be necessary to have several useful circuits, such as LED monitors, 7-segment displays, debounced switches, etc.

LED Monitors

A *light-emitting diode* (*LED*) *monitor* is a monitor in which the LED is lit for a logic 1 state and unlit for a logic 0 state. A circuit for use with TTL and CMOS devices is shown in Fig. 2-4. When this monitor is used, the schematic symbol of Fig. 2-5 will be shown in the *Schematic Diagram of Circuit* section of the experiment.

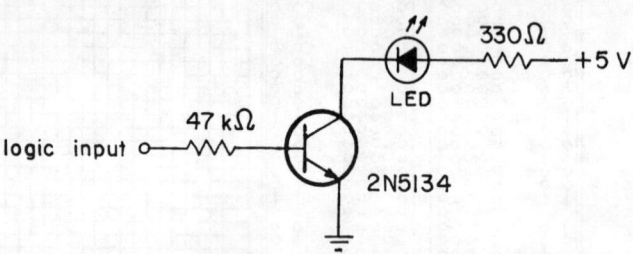

Fig. 2-4. LED monitor.

Fig. 2-5. LED monitor schematic symbol.

16

Fig. 2-6. TTL logic switch.

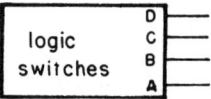

Logic Switches

A *logic switch* is a mechanical spdt switch that applies either a logic 0 or a logic 1 state at its output terminal. For TTL devices, the circuit of Fig. 2-6 can be used. When using CMOS devices, the 1-kΩ resistors should be replaced by 2.2-kΩ resistors.

Usually, you will need logic switches in groups of four, representing the inputs of a 4-bit bcd number. Fig. 2-7 shows the schematic diagram that will be used to represent a series of logic switches.

Fig. 2-7. Logic switch schematic symbol.

logic switches	D C B A

Debounced Switches (Pulsers)

In a *mechanical* logic switch, contact bounce (i.e., the uncontrolled making and breaking of contact when the switch contacts are opened or closed) is a common occurrence. In most digital applications, it is extremely important that the output from a switch be bounce-free, or *debounced*. When experimenting with TTL devices, the circuit

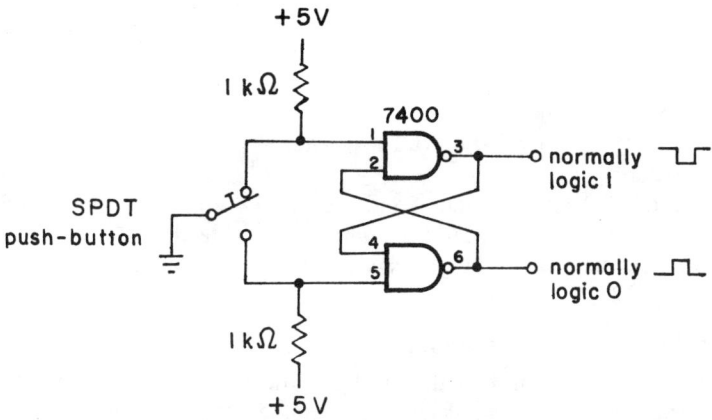

Fig. 2-8. TTL pulser.

17

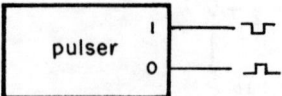

Fig. 2-9. Pulser schematic symbol.

shown in Fig. 2-8 is useful as a push-button debounced switch or pulser. Using a pair of 7400 NAND gates, the pulser has complementary logic 0 and logic 1 outputs. For experiments with CMOS devices, use either a 74C00 or a 4011 quad NAND gate, and replace the 1-kΩ resistors with 2.2-kΩ resistors. Fig. 2-9 shows the schematic symbol that will be used to represent a pulser.

7-Segment LED Displays

To easily determine the state of binary counters or 4-bit bcd input data, we can use a *7-segment LED display*. Using a 7447 bcd-to-7 segment decoder/driver and a MAN-7 *common-anode* display, we can use the circuit shown in Fig. 2-10 to decode bcd data from TTL devices.

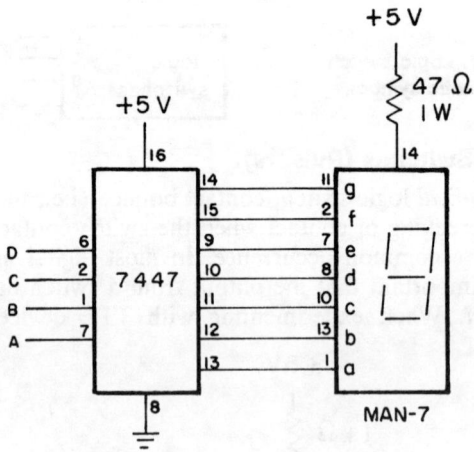

Fig. 2-10. TTL 7-segment LED display.

For CMOS circuits, the circuit shown in Fig. 2-11 uses a 4511 decoder/driver and a MAN-3 *common-cathode* display. The 7-segment LED display will be represented by the schematic symbol shown in Fig. 2-12.

Stable Frequency Reference

In order to accurately demonstrate the operation of either fixed or programmable divide-by-N counters, as well as frequency synthesizers, a stable square-wave frequency reference will be needed.

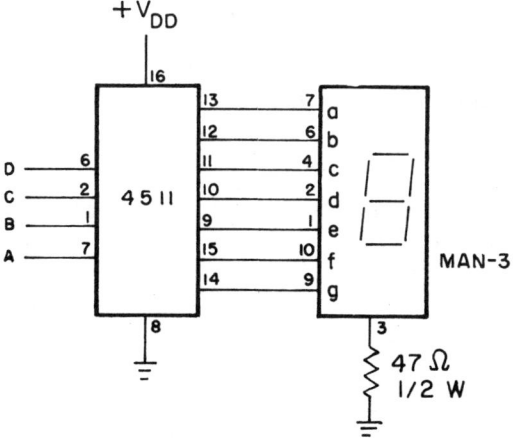

Fig. 2-11. CMOS 7-segment LED display.

Fig. 2-12. Schematic symbol for 7-segment LED display.

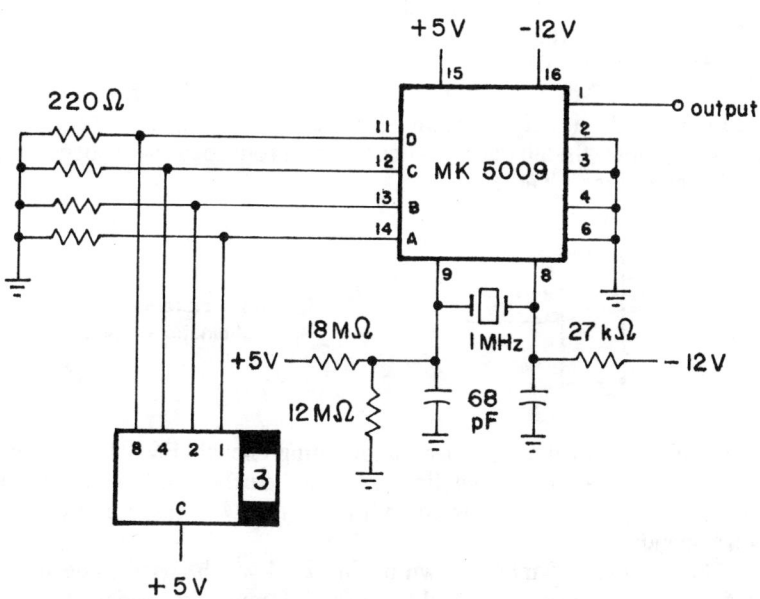

Fig. 2-13. Experimenter-type stable frequency reference.

19

Such a device has its output frequency controlled by a quartz crystal, similar to those used for electronic wristwatches. Chapter 6 discusses several TTL and CMOS circuits; however, a more versatile experimenter-type circuit is shown in Fig. 2-13 and can be used with both TTL and CMOS devices.

Table 2-1. Input Code for Programming the MK5009

BCD Inputs				Divisor	Output Frequency
D	C	B	A		
0	0	0	0	$\div 10^0$	1 MHz
0	0	0	1	$\div 10^1$	100 kHz
0	0	1	0	$\div 10^2$	10 kHz
0	0	1	1	$\div 10^3$	1 kHz
0	1	0	0	$\div 10^4$	100 Hz
0	1	0	1	$\div 10^5$	10 Hz
0	1	1	0	$\div 10^6$	1 Hz
0	1	1	1	$\div 10^7$	0.1 Hz
1	0	0	0	$\div 10^8$	0.01 Hz

Using an MK5009 integrated-circuit oscillator/divider (MOSTEK Corp.), this circuit is capable of generating square-wave outputs from 1 MHz down to 0.01 Hz in 1-decade steps with a single crystal. Programming the MK5009 to generate the various output frequencies is accomplished with a thumbwheel switch (see Chapter 6 for a discussion about thumbwheel switches) to produce a 4-bit bcd input code according to Table 2-1.

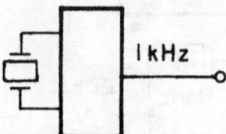

Fig. 2-14. Reference frequency schematic symbol.

In addition to the SK-10 breadboarding socket, E&L Instruments manufactures an extensive line of breadboarding aids representing most of the circuits discussed in this chapter. These are presented in Appendix C.

The schematic symbol shown in Fig. 2-14 will be used to designate a frequency reference. In addition, the reference frequency will also be indicated (e.g., 1 kHz).

COMPONENTS

The following is a list of all the various components needed to perform all the experiments given in this book.

Fixed Resistors

1—220 Ω	5—10 kΩ	1—22 kΩ
3—560 Ω	1—15 kΩ	1—27 kΩ
1—4.7 kΩ	1—18 kΩ	2—100 kΩ
		1—560 kΩ

Potentiometers

1—1 kΩ   1—10 kΩ

Capacitors

2—0.001 μF	3—0.047 μF	1—5 μF
1—0.01 μF	3—0.1 μF	1—50 μF
4—0.022 μF	1—0.33 μF	

TTL Integrated Circuits

- 1—7402 quad 2-input NOR gate
- 1—7404 hex inverter
- 1—7442 bcd to 1-of-10 decoder
- 2—7474 dual D-type edge-triggered flip-flop
- 1—7486 quad exclusive-OR gate
- 1—7490 decade counter
- 1—7492 divide-by-12 counter
- 1—74192 presettable up/down decade counter
- 1—MC4024 dual voltage-controlled multivibrator (Motorola)
- 1—MC4044 phase-frequency detector (Motorola)

CMOS Integrated Circuits

- 1—4001 quad 2-input NOR gate
- 1—4017 decade counter
- 1—4046 phase-locked loop
- 1—HCTR 0320 phase detector/programmable divider (Hughes)

Linear Integrated Circuits

- 2—555 timer (8-pin DIP)
- 1—565 phase-locked loop (14-pin DIP)
- 1—567 phase-locked loop/tone decoder (8-pin DIP)
- 1—741 operational amplifier (8-pin DIP)

Solid-State Devices
   1—1N914 diode
   1—2N2222 npn transistor

Miscellaneous
   1—8-Ω speaker

The following is a list of the various mail-order sources for the integrated circuits and other components for the experiments, as well as other components for the experimenter.

Semiconductor Technology (for the HCTR 0320 chip)
124-14 22nd Ave.
College Point, NY 11356

Jade Computer Products
5351 West 144th St.
Lawndale, CA 90260

Quest Electronics
P.O. Box 4430M
Santa Clara, CA 95054

Integrated Circuits Unlimited
7889 Clairemont Mesa Blvd.
San Diego, CA 92111

Poly Paks
P.O. Box 942A
Lynnfield, MA 01940

Solid State Sales
P.O. Box 74A
Somerville, MA 02143

James Electronics
1021-A Howard Ave.
San Carlos, CA 94070

Godbout Electronics
P.O. Box 2355
Oakland, CA 94614

Digi-Key Corp.
P.O. Box 677
Thief River Falls, MN 56701

S.D. Sales Co.
P.O. Box 28810
Dallas, TX 75228

Astral Electronics Corp.
321 Pennsylvania Ave., P.O. Box 707
Linden, NJ 07036

CHAPTER 3

The Phase Detector

INTRODUCTION

In this chapter, several types of commonly used discrete phase detectors are discussed, each with their strong and weak points. Some types of phase detectors such as the exclusive-OR and edge-triggered types can be built using simple logic elements. On the other hand, the MC4044 is a complex device, available as a monolithic integrated circuit.

OBJECTIVES

At the completion of this chapter, you will be able to do the following:

- Define the terms *phase* and *phase difference.*
- Determine the phase difference of two signals having the same frequency in terms of time units, electrical degrees, and radians.
- Describe the input and output characteristics of the following types of phase detectors:
  exclusive-OR
  edge-triggered
  MC4044 integrated circuit
- Determine the conversion gain (K_ϕ) for several types of phase detectors.

PHASE

The term *phase* basically refers to the interval between the time when one event occurs and the time when a second, but related event occurs. The event that occurs first is said to *lead,* while the second event is said to *lag* the first. As shown in Fig. 3-1, signal A leads signal B. In other words, the point on signal B (i.e., t_2) takes place t units of time *later* than the corresponding point on signal A (t_1), assuming that both periodic signals are of the same frequency.

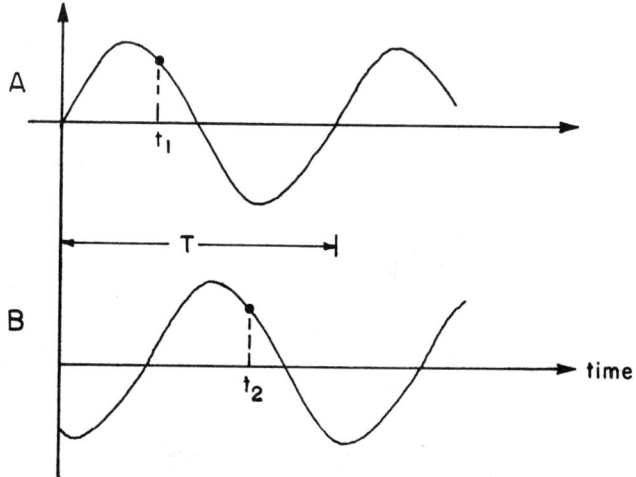

Fig. 3-1. The time (phase) difference of two signals.

The *phase difference* is the time, expressed in electrical degrees, in which one signal leads or lags another, and is usually less than one cycle. As before, the phase difference between two signals can be simply expressed in ordinary time units, as illustrated in Fig. 3-1, but it is more conveniently expressed in *degrees* (sometimes called *electrical degrees*), and given the symbol $\Delta\phi$. Since each cycle of either signal A or signal B occupies exactly the same amount of time, *using the cycle as the time unit makes the specification of the phase difference independent of the frequency of the signal.*

Since, by definition, one complete cycle is equal to 360°, the phase difference between signals A and B is the fraction of one cycle expressed in degrees, so that

$$\Delta\phi = \frac{t_2 - t_1}{T} \times 360° \qquad \text{(Eq. 3-1)}$$

Example

Determine the phase difference between the two periodic rectangular waveforms shown in Fig. 3-2. It should first be noted that waveform B goes positive 0.4 second later than the corresponding point on waveform A. Since the total time to complete one cycle is 1.5 seconds, the phase difference is then

$$\Delta\phi = \frac{0.4}{1.5}(360°)$$
$$= 96°$$

so that waveform A *leads* waveform B by 96°. However, it is also correct to say that B *lags* A by 96°.

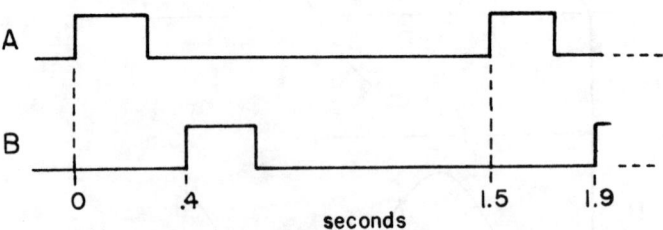

Fig. 3-2. Example illustration for determining the phase difference between two signals.

THE PHASE DETECTOR

All phase-locked loop systems use a circuit called a *phase detector*, or *phase comparator*. The phase detector generates an average, or dc, output voltage that is proportional to the phase difference between the input of the phase-locked loop and the vco. The output voltage is often referred to as the *error voltage*. The factor that converts phase difference into voltage is called the *phase-detector conversion gain*, so that

$$V_o = K_\phi \Delta\phi \qquad \text{(Eq. 3-2)}$$

where,
V_o is the average output voltage of the phase detector in volts,
K_ϕ is the phase detector conversion gain in volts/radian,
$\Delta\phi$ is the input phase difference in radians.

In the previous section, phase difference was expressed in terms of degrees. When working with phase-locked loops, it is customary to express this difference in terms of *radians*. In terms of electrical degrees, 1 radian is equivalent to $180°/\pi$, or 57.3°. For the previous example, a phase difference of 96° is equivalent to 96/57.3, or 1.68 radians.

One basic difference between analog- and digital-type phase-locked loops is with respect to the type of phase detector that is used. In general, analog phase-locked loops use a double-balanced mixer, while digital phase-locked loops use either an exclusive-OR, or some type of edge-triggered phase detector. Since the majority of the discrete phase detectors in use are of the digital type, the types discussed in this chapter will be digital. However, monolithic phase-locked-loop integrated circuits frequently use an analog phase detector, and are briefly discussed in Chapter 7.

THE EXCLUSIVE-OR PHASE DETECTOR

The exclusive-OR type of phase detector uses, as its name implies, an exclusive-OR logic gate, as shown symbolically in Fig. 3-3. For this 2-input gate we have the truth table of Table 3-1.

Table 3-1. Truth Table for Fig. 3-3

Inputs		Output
A	B	Q
0	0	0
1	0	1
0	1	1
1	1	0

Fig. 3-3. The exclusive-OR logic gate.

Because of this truth table, exclusive-OR-type phase detectors are used for input and vco waveforms that have a 50% duty cycle; i.e., symmetrical. As an illustration, Fig. 3-4 shows the input signal leading the vco signal by $\pi/4$ radians, or 45°. In Fig. 3-5, the vco signal lags

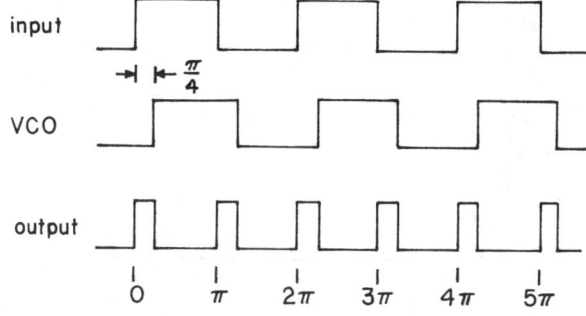

Fig. 3-4. Input signal leads vco signal by $\pi/4$, or 45°.

the input by $\pi/2$, or 90°. In either case, the output of the exclusive-OR gate is now a rectangular pulse train, the frequency of which is twice that of the input. More importantly, the output pulse width depends solely on the phase difference of the detector's two inputs.

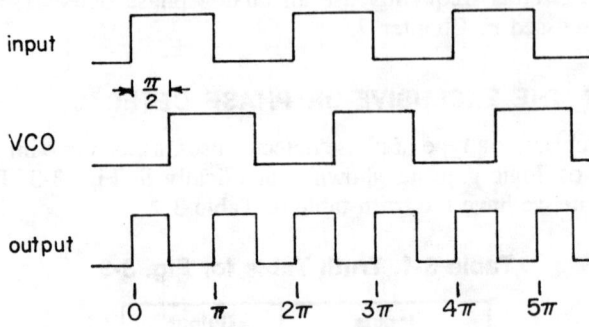

Fig. 3-5. Input signal leads vco signal by $\pi/2$, or 90°.

Consequently, the average, or *dc,* output voltage will also depend on the phase difference and is related to the duty cycle of the phase detector's output so that

$$V_o(dc) = V_p \times D \qquad \text{(Eq. 3-3)}$$

where,
V_p is the maximum output voltage (logic 1),
D is the duty cycle.

The *duty cycle* of a periodic rectangular pulse train is defined as the ratio of the duration that the waveform is at logic 1 to the duration of one complete cycle. For the output waveform shown in Fig. 3-4, the duty cycle is the ratio of $\pi/4$ to π radians, or 0.25 (25%). For the output shown in Fig. 3-5, it is 0.5 (50%).

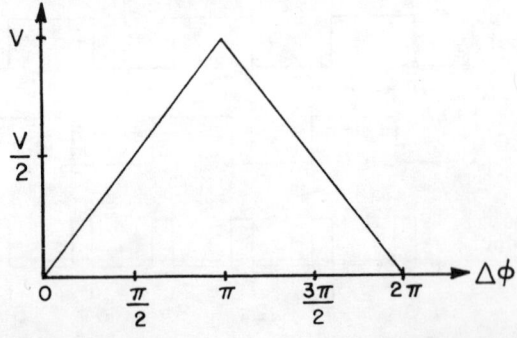

Fig. 3-6. Exclusive-OR phase detector input/output characteristic.

By plotting the average output voltage of the exclusive-OR phase detector as a function of the phase difference of its two inputs, a triangular characteristic results, as shown in Fig. 3-6.

As the phase difference increases from 0 toward π, or 3.14 radians, the average voltage of the output waveform approaches a maximum value (slightly less than the supply voltage for standard logic families) at $\Delta\phi = \pi$ radians, or 180°. The slope of the line over this range is the *phase detector conversion gain* (K_ϕ) and is expressed in

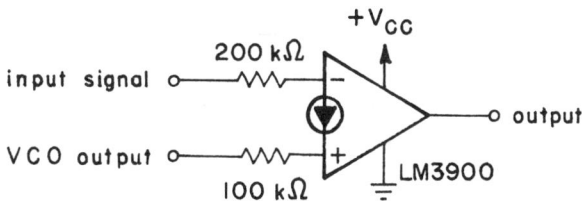

Fig. 3-7. Exclusive-OR logic element using a type LM3900 Norton op amp.

units of volts/radian. As the phase difference increases from π toward 2π radians, the average output voltage decreases linearly, and the slope is also equal to the phase detector conversion gain. The TTL exclusive-OR gate is the 7486 integrated circuit, while the CMOS equivalent is the 4030 or 74C86. On the other hand, an exclusive-OR logic element can be made using a type LM3900 Norton operational amplifier, as shown in the circuit of Fig. 3-7.

EDGE-TRIGGERED PHASE DETECTORS

A second type of digital phase detector is the *edge-triggered* type. One of the simplest types of edge-triggered detectors is the *set-reset,*

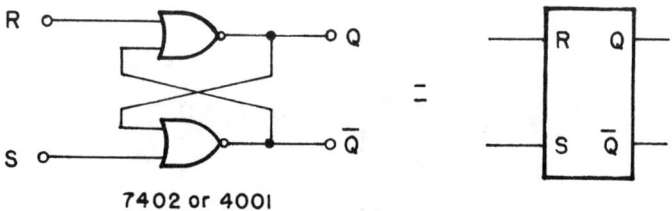

Fig. 3-8. The R-S flip-flop as an edge-triggered phase detector.

or *R-S flip-flop*. As shown in Fig. 3-8, the R-S flip-flop can be made from a pair of cross-coupled NOR gates.

The two basic rules governing the operation of the R-S flip-flop as an edge-triggered phase detector are as follows:

1. If the *set* or *S* input is at logic 1, the Q output goes to or stays at logic 1, while the $\overline{Q}$ output goes to or stays at logic 0 (ground).
2. If the *reset* or *R* input is at logic 1, the Q output goes to or stays at logic 0, while the $\overline{Q}$ output goes to or stays at logic 1.

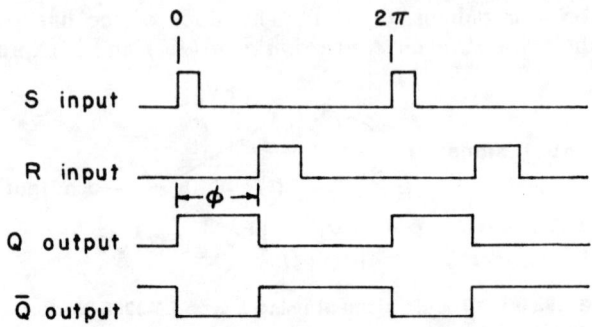

Fig. 3-9. Input/output waveforms for the R-S flip-flop edge-triggered phase detector.

As shown in the timing diagram of Fig. 3-9, the NOR gate R-S flip-flop is triggered on the *positive leading edge* of the two inputs. For the R-S flip-flop, as well as other types of edge-triggered detectors, the input pulses are usually of short duration rather than the symmetrical 50% duty-cycle pulses associated with the exclusive-OR detector. Similar to the exclusive-OR detector, as the phase difference between the input and vco increases, the average output voltage of the edge-triggered detector increases proportionately. By plotting the average output voltage as a function of the phase difference between th S and R inputs, a sawtooth-shaped characteristic is obtained, as shown in Fig. 3-10. Consequently, the edge-triggered–type detector

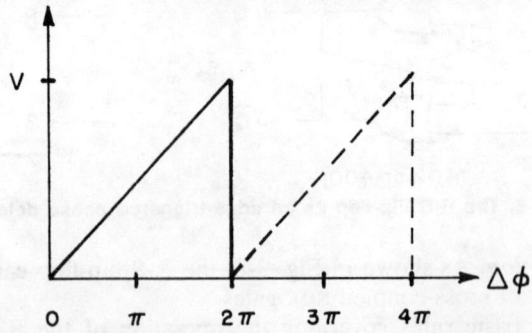

Fig. 3-10. Edge-triggered phase detector input/output characteristic.

has twice the linear range as the double-valued triangular curve of the exclusive-OR detector (Fig. 3-6). In addition, the edge-triggered detector will possess significantly better capture, tracking, and locking characteristics than the exclusive-OR detector.

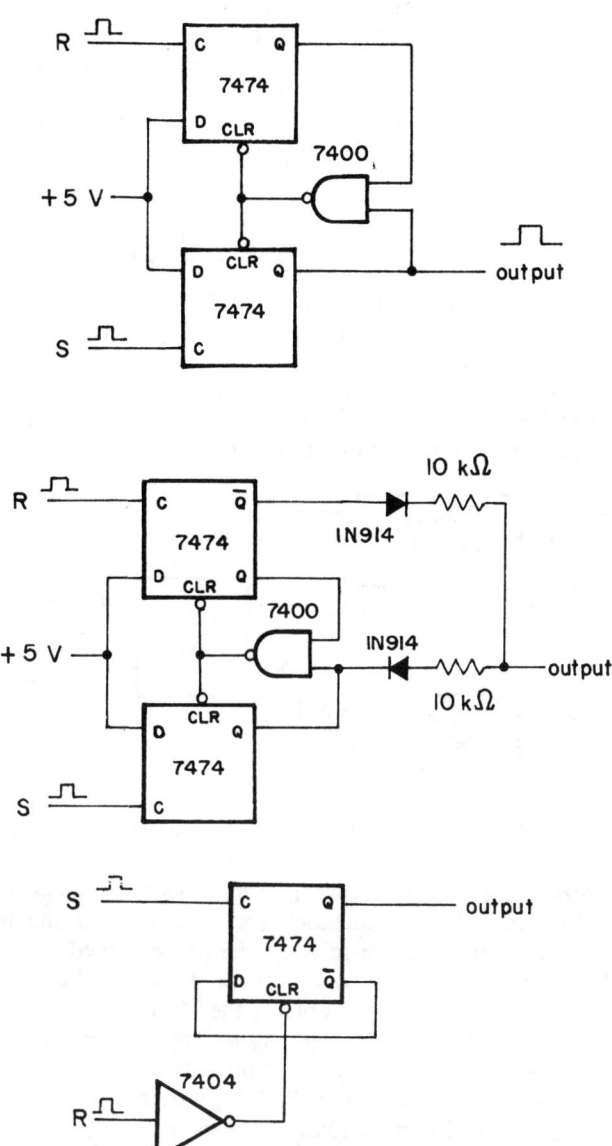

Fig. 3-11. Edge-triggered phase-detector circuits using the D-type flip-flop.

In addition to the NOR gate R-S flip-flop circuit of Fig. 3-8, there are several other edge-triggered circuits in use, primarily built around the D-type flip-flop, as shown in Fig. 3-11.

Both the exclusive-OR and the edge-triggered phase detector are sensitive to harmonic multiples of the incoming signal. Therefore, the phase-locked loop tends to lock onto these harmonics. In addition, both types are sensitive to changing duty cycles of the phase detector's two inputs. If the duty cycle of either input of the exclusive-OR detector is not 0.5 (50%), an extraneous error results. For the edge-triggered detector, if the R input is at logic 1 when the S input is also logic 1, the detector will not function properly.

THE MC4044 PHASE DETECTOR

The MC4044 integrated circuit, manufactured by Motorola, is a monolithic 14-pin DIP phase detector which avoids both the harmonic sensitivity and the duty-cycle problems associated with the exclusive-OR and edge-triggered detectors. As shown in the block diagram of Fig. 3-12, the MC4044 consists of two digital phase detectors, a charge pump, and an amplifier.

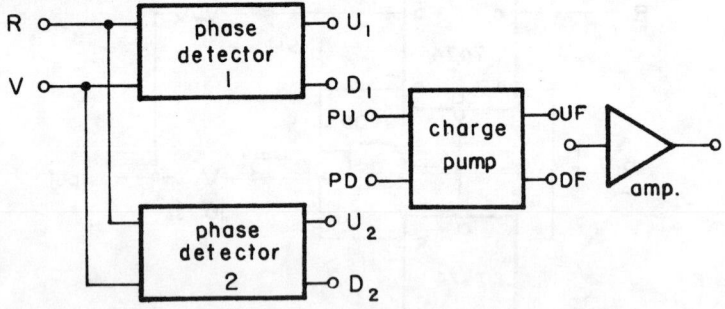

Fig. 3-12. Block diagram of the MC4044 phase detector.

Phase detector No. 1, shown in Fig. 3-13A, is a negative edge-triggered circuit which is intended for systems requiring both zero frequency and phase difference when the phase-locked loop is phase-locked. For a given phase condition, only one of the two outputs, U_1 or D_1, is active. As an example, if the V or *variable* input signal, such as the vco, *lags* the R or *reference* input (the input signal to the phase-locked loop), a signal is present at output U_1 (up), as shown in Fig. 3-13B. However, if R lags V, the output appears at D_1 (down), as seen in Fig. 3-13C.

The input/output transfer characteristic curve of phase detector No. 1 is a sawtooth similar to that of the edge-triggered type, except

HOWARD W. SAMS & COMPANY

Bookmark

DEAR VALUED CUSTOMER:

Howard W. Sams & Company is dedicated to bringing you timely and authoritative books for your personal and professional library. Our goal is to provide you with excellent technical books written by the most qualified authors. You can assist us in this endeavor by checking the box next to your particular areas of interest.

We appreciate your comments and will use the information to provide you with a more comprehensive selection of titles.

Thank you,

Vice President, Book Publishing
Howard W. Sams & Company

COMPUTER TITLES:

Hardware
- ☐ Apple 140   ☐ Macintosh 101
- ☐ Commodore 110
- ☐ IBM & Compatibles 114

Business Applications
- ☐ Word Processing J01
- ☐ Data Base J04
- ☐ Spreadsheets J02

Operating Systems
- ☐ MS-DOS K05   ☐ OS/2 K10
- ☐ CP/M K01   ☐ UNIX K03

Programming Languages
- ☐ C L03   ☐ Pascal L05
- ☐ Prolog L12   ☐ Assembly L01
- ☐ BASIC L02   ☐ HyperTalk L14

Troubleshooting & Repair
- ☐ Computers S05
- ☐ Peripherals S10

Other
- ☐ Communications/Networking M03
- ☐ AI/Expert Systems T18

ELECTRONICS TITLES:
- ☐ Amateur Radio T01
- ☐ Audio T03
- ☐ Basic Electronics T20
- ☐ Basic Electricity T21
- ☐ Electronics Design T12
- ☐ Electronics Projects T04
- ☐ Satellites T09

- ☐ Instrumentation T05
- ☐ Digital Electronics T11

Troubleshooting & Repair
- ☐ Audio S11   ☐ Television S04
- ☐ VCR S01   ☐ Compact Disc S02
- ☐ Automotive S06
- ☐ Microwave Oven S03

Other interests or comments: _____
_____
_____

Name _____
Title _____
Company _____
Address _____
City _____
State/Zip _____
Daytime Telephone No. _____

A Division of Macmillan, Inc.
4300 West 62nd Street
Indianapolis, Indiana 46268

21545

Bookmark

HOWARD W. SAMS & COMPANY

NO POSTAGE
NECESSARY
IF MAILED
IN THE
UNITED STATES

BUSINESS REPLY CARD

FIRST CLASS   PERMIT NO. 1076   INDIANAPOLIS, IND.

POSTAGE WILL BE PAID BY ADDRESSEE

HOWARD W. SAMS & CO.
ATTN: Public Relations Department
P.O. BOX 7092
Indianapolis, IN 46209-9921

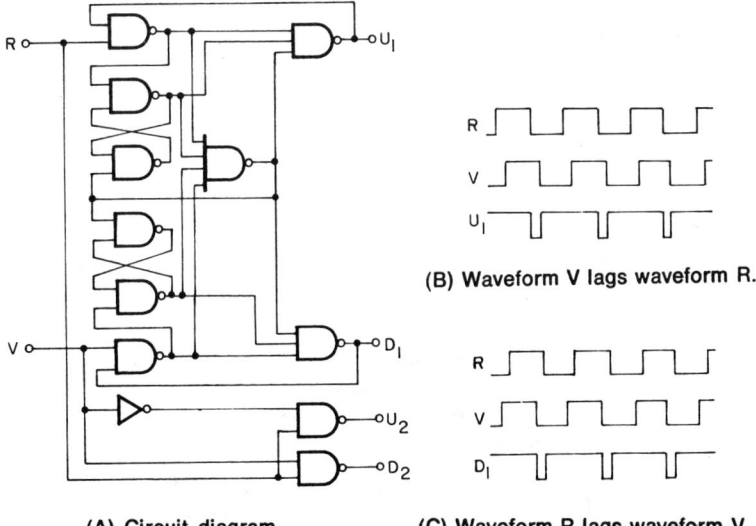

(A) Circuit diagram. (C) Waveform R lags waveform V.

Fig. 3-13. Circuit diagram and input/output waveforms for the MC4044 phase detector.

that the MC4044 has an even wider linear range of 4π radians (Fig. 3-14). Typically, the conversion gain is 0.12 V/rad. Outputs U_1 and D_1 are then connected to the PU and PD inputs, respectively, of the device's charge pump, whose output varies approximately from +0.75 to +2.25 volts as the phase difference between R and V varies from -2π to $+2\pi$ radians.

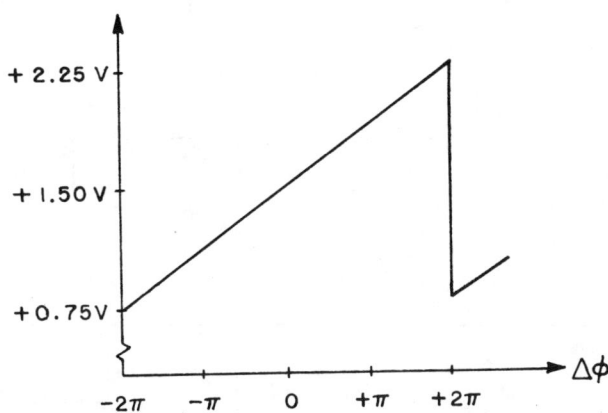

Fig. 3-14. MC4044 input/output characteristic for phase detector No. 1.

33

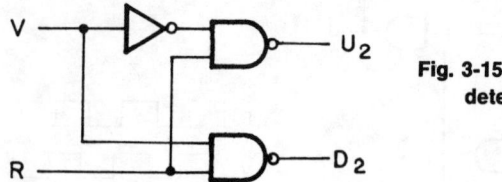

Fig. 3-15. MC4044 phase detector No. 2.

As shown in Fig. 3-15, phase detector No. 2 consists only of combinatorial logic, so that its truth table can be represented as shown in Table 3-2.

Table 3-2. Truth Table for Fig. 3-15

Inputs		Outputs	
R	V	U_2	D_2
0	0	1	1
1	0	0	1
0	1	1	1
1	1	1	0

Waveforms showing the operation of phase detector No. 2 when phase detector No. 1 is being used as part of the phase-locked-loop system are shown in Fig. 3-16. When the loop is phase-locked, output U_2 remains at logic 1. If the loop drifts out of phase-lock, a negative pulse whose width is proportional to the amount of drift

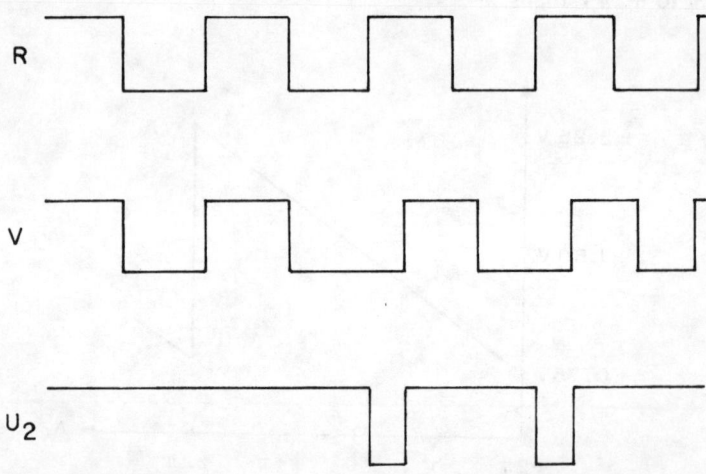

Fig. 3-16. Waveforms for MC4044 phase detector No. 2.

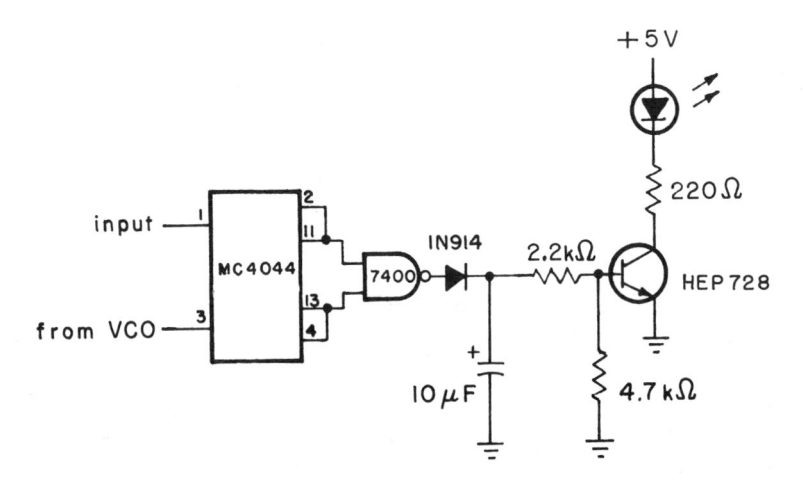

Fig. 3-17. MC4044 "loss-of-lock" indicator.

appears at U_2. Consequently, this feature can be used to make a "loss-of-lock" indicator, using the circuit of Fig. 3-17.

AN INTRODUCTION TO THE EXPERIMENTS

The following experiments are designed to demonstrate the measurement of phase difference, average output voltage, and the input/output characteristics of different types of phase detectors. For each type you will be able to determine the conversion gain and the operating range. The experiments that you will perform can be summarized as follows:

Experiment No. *Purpose*

1   Measure the phase difference between two square-wave signals having the same frequency.

2   Determine the operation of an exclusive-OR logic gate by constructing a truth table.

3   Determine the input/output characteristics of an exclusive-OR phase detector.

4   Determine the operation of an R-S edge-triggered flip-flop made from two NOR gates.

35

5   Determine the input/output characteristics of an edge-triggered phase detector using a D-type flip-flop.

6   Determine the input/output characteristics of the Motorola MC4044 integrated-circuit phase detector.

EXPERIMENT NO. 1

Purpose

The purpose of this experiment is to measure the phase difference between two square-wave signals having the same frequency. To generate the various phase differences, we shall use a circuit built around a pair of 7474 dual D-type flip-flops.

Pin Configuration of Integrated-Circuit Chip (Fig. 3-18)

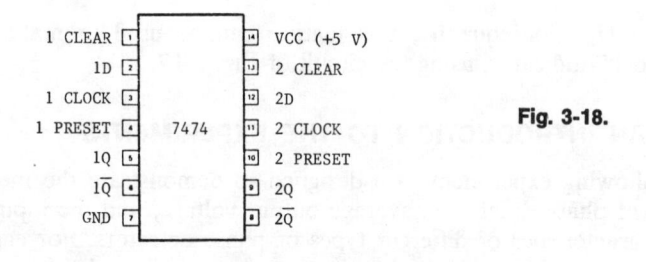

```
1 CLEAR  [1]          [14] VCC (+5 V)
1 D      [2]          [13] 2 CLEAR
1 CLOCK  [3]   7474   [12] 2D
1 PRESET [4]          [11] 2 CLOCK
1Q       [5]          [10] 2 PRESET
1Q̄       [6]          [9]  2Q
GND      [7]          [8]  2Q̄
```

Fig. 3-18.

Schematic Diagram of Circuit (Fig. 3-19)

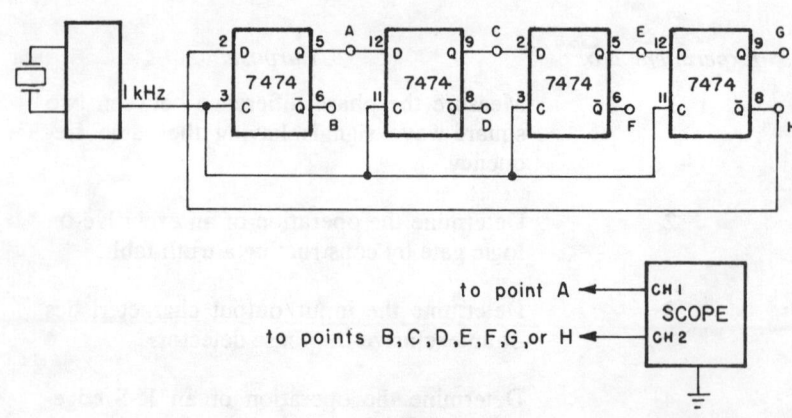

to point A → CH1 SCOPE
to points B, C, D, E, F, G, or H → CH2

Fig. 3-19.

Design Basics

$$\Delta\phi = \frac{t}{T} \times 360°$$
Phase difference:

Step 1

Set your oscilloscope for the following settings:

- Channels 1 & 2: 5 V/division
- Time base: 0.5 ms/division
- Dc coupling
- Trigger Channel 1

Step 2

Wire the circuit shown in the schematic diagram, making sure that you have properly connected the power-supply connections to the 7474 integrated circuits; pin 7 to ground and pin 14 to +5 V.

Step 3

After you have made sure your circuit is wired correctly, apply power to the breadboard and then connect the frequency source to the circuit.

Step 4

Now connect the channel 1 probe of the oscilloscope to point A (pin 5 of the first flip-flop) and channel 2 probe also to point A. Since both probes are connected at the same point, you should observe two square-wave signals that vary exactly with time. Position the channel 1 trace above channel 2. When making phase-difference measurements, we shall use the convention that channel 1 is the *reference signal* while channel 2 is the *variable signal*.

Step 5

Now adjust the oscilloscope's time-base calibration control until the period of both traces is exactly 10 horizontal divisions wide, as shown in the oscilloscope display of Fig. 3-20.

Step 6

Now connect the channel 2 probe to point B shown on the schematic diagram. You should see that the waveform of channel 2 is *inverted* with respect to that of channel 1 (point A). The waveform of channel 2 goes *high* exactly 5 divisions *later* than that of channel 1. Since the total period is 10 divisions (T), and the waveform at point B occurs 5 divisions later than the waveform at point A (t), the phase difference is

$$\Delta\phi = \frac{t}{T} \times 360°$$
$$= (5/10)(360°)$$
$$= 180°$$

so that the waveform at point A *leads* the waveform at point B by 180°.

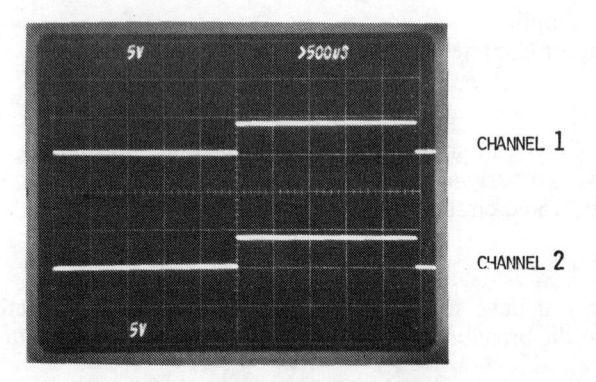

Fig. 3-20. Oscilloscope display.

Step 7

Now connect the channel 2 probe to point C shown on the schematic diagram. You should observe that the waveform of channel 2 goes *high* 1.25 divisions later than that of channel 1, so the phase difference is

$$\Delta\phi = (1.25/10)(360°)$$
$$= 45°$$

so that point A leads point B by 45°.

Step 8

Now continue to measure the phase difference at points D, E. F, G, and H with respect to point A, recording your results in Table 3-3. If you have done everything correctly and accurately, your results should be as shown in Table 3-4.
If not, repeat the entire experiment.
Save this circuit on a portion of your breadboard as it will be needed for Experiment No. 3.

Table 3-3.

Channel 1	Channel 2	Δ Divisions	Δϕ (A leads B)
A	A	0	0°
A	B	5	180°
A	C	1.25	45°
A	D		
A	E		
A	F		
A	G		
A	H		

Table 3-4.

Channel 1	Channel 2	Δ Divisions	Δϕ (A leads B)
A	A	0	0°
A	B	5	180°
A	C	1.25	45°
A	D	6.25	225°
A	E	2.5	90°
A	F	7.5	270°
A	G	3.75	135°
A	H	8.75	315°

EXPERIMENT NO. 2

Purpose

The purpose of this experiment is to determine the operation of the 7486 TTL exclusive-OR gate by constructing its truth table.

Pin Configuration of Integrated-Circuit Chip (Fig. 3-21)

Fig. 3-21.

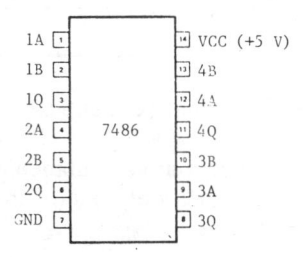

```
1A  [1]        [14] VCC (+5 V)
1B  [2]        [13] 4B
1Q  [3]        [12] 4A
2A  [4]  7486  [11] 4Q
2B  [5]        [10] 3B
2Q  [6]        [9]  3A
GND [7]        [8]  3Q
```

39

Schematic Diagram of Circuit (Fig. 3-22)

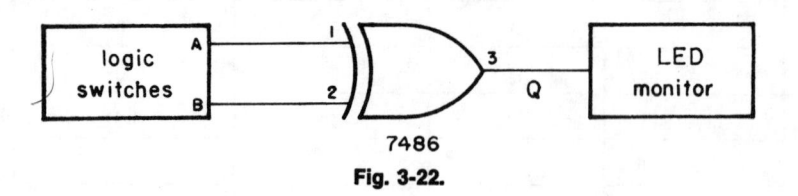

```
                    1 \
  logic      A  ─────  \
  switches      ─────   )D──── 3 ────  LED
             B  2   /         Q        monitor
                   /
                 7486
```

Fig. 3-22.

Step 1
Wire the circuit shown in the schematic diagram. Don't forget the power connections at pins 7 (ground) and 14 (+5 V) of the 7486 chip, as these are normally implied when omitted from the schematic diagram.

Step 2
Apply power to the breadboard and set both logic switches A and B to logic 0. The LED monitor should be *unlit,* indicating that the Q output of the exclusive-OR gate is logic 0.

Step 3
Vary the settings of the logic switches and complete Table 3-5, called a *truth table,* based on the output you observe on the LED monitor (unlit is logic 0, and lit is logic 1).

Table 3-5.

Inputs		Output
A	B	Q
0	0	0
1	0	
0	1	
1	1	

Your results should be identical to Table 3-6 for the 7486 exclusive-OR gate.

When *both* inputs are simultaneously at logic 0 *or* logic 1, the output is at logic 0. When only one of the two inputs is at logic 1, then the output is at logic 1.

Table 3-6.

Inputs		Output
A	B	Q
0	0	0
1	0	1
0	1	1
1	1	0

EXPERIMENT NO. 3

Purpose

The purpose of this experiment is to determine the characteristics of the exclusive-OR phase detector, using the phase generator of Experiment No. 1, and a 7486 TTL exclusive-OR gate.

Pin Configuration of Integrated-Circuit Chip (Fig. 3-23)

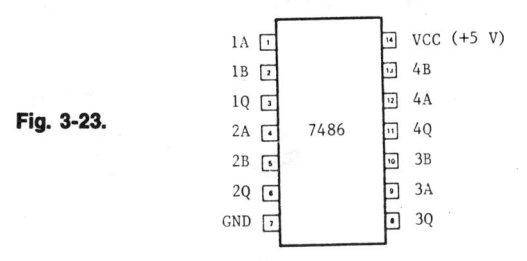

Fig. 3-23.

```
1A  [1]            [14] VCC (+5 V)
1B  [2]            [13] 4B
1Q  [3]            [12] 4A
2A  [4]   7486     [11] 4Q
2B  [5]            [10] 3B
2Q  [6]            [9]  3A
GND [7]            [8]  3Q
```

Schematic Diagram of Circuit (Fig. 3-24)

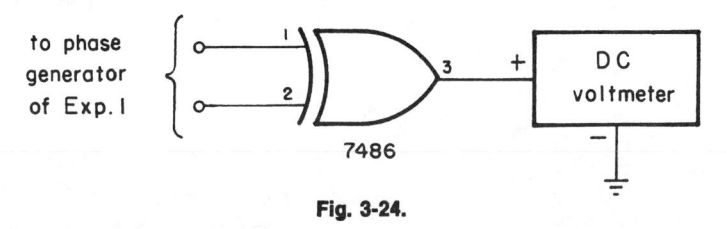

Fig. 3-24.

Step 1

Wire the circuit shown in the schematic diagram. If you have not already done so, wire the phase-generator circuit given in Experiment No. 1. Then apply power to the breadboard.

Step 2

Connect pin 1 of the 7486 to point A of the phase-generator circuit, and pin 2 of the 7486 also to point A. From Experiment No. 1

we already know that the phase difference will be 0°. Using the dc voltmeter, measure the voltage at pin 3 of the 7486 and record your result:

$$V_o(\Delta\phi = 0°) = \underline{} \text{ volts}$$

You should measure a value of about 0.075 volt.

Step 3

Now connect pin 2 of the 7486 exclusive-OR gate to point B of the phase-generator circuit. Now the phase difference will be 180°. Measure the output voltage at pin 3 and record your result:

$$V_o(\Delta\phi = 180°) = \underline{} \text{ volts}$$

Step 4

Continue the experiment by measuring the output voltage of the exclusive-OR gate by connecting pin 2 to the remaining points on the phase-generator circuit, recording your results in Table 3-7.

Step 5

Now plot your results from Steps 2, 3, and 4 on the blank graph (Fig. 3-25) provided for this purpose. If you have done this experiment correctly, you should have plotted a triangle-shaped curve

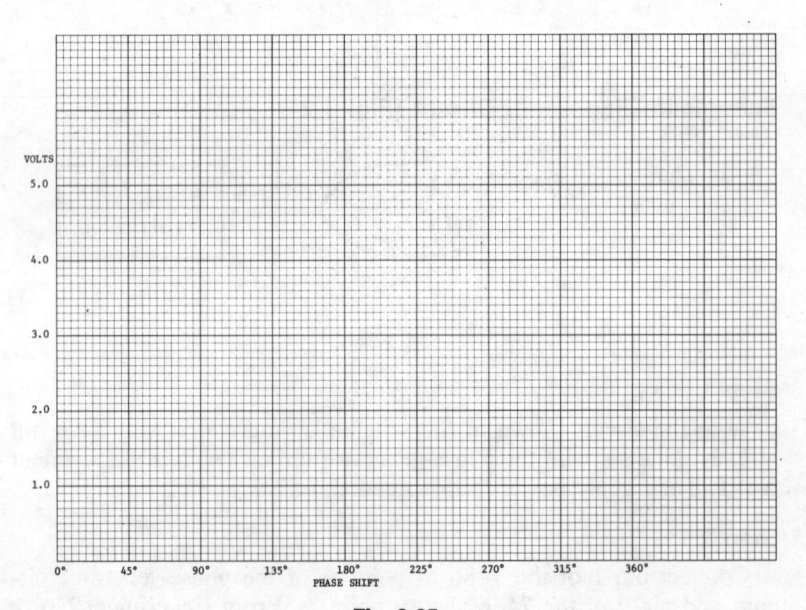

Fig. 3-25.

Table 3-7.

Pin 2 Connected to	Δφ	Output Voltage
Point C	45°	
Point D	225°	
Point E	90°	
Point F	270°	
Point G	135°	
Point H	315°	

which is symmetrical about 180° (see Fig. 3-6). As the phase difference increases from 0 toward 180° (or 0 to π radians), the output voltage of the exclusive-OR phase detector increases linearly and reaches its maximum voltage at 180°.

Step 6

Now compute the conversion gain (K_ϕ) for this exclusive-OR phase detector by taking the slope of the line. To do this, subtract the voltage measured at 0° phase difference (Step 2) from the voltage measured at 180° phase difference (Step 3). Then divide this difference by π (i.e., 3.14) radians:

$$K_\phi(\text{exclusive-OR}) = \frac{V_{o(180°)} - V_{o(0°)}}{\pi}$$

$$= \underline{\qquad} \text{ volts/radian}$$

The output voltage should vary from about 0.075 to 3.84 volts, giving a phase-detector conversion gain of 1.20 volts/radian.

EXPERIMENT NO. 4

Purpose

The purpose of this experiment is to demonstrate the operation of an R-S edge-triggered flip-flop, made from two 7402 NOR gates.

Pin Configuration of Integrated-Circuit Chip (Fig. 3-26)

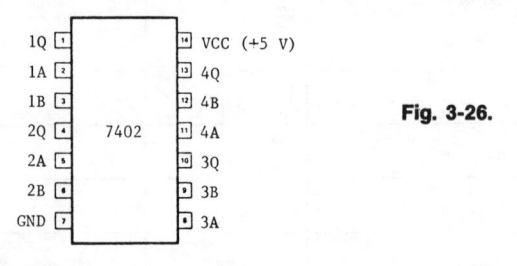

```
1Q  [1]        [14] VCC (+5 V)
1A  [2]        [13] 4Q
1B  [3]        [12] 4B
2Q  [4] 7402   [11] 4A
2A  [5]        [10] 3Q
2B  [6]        [9]  3B
GND [7]        [8]  3A
```

Fig. 3-26.

Schematic Diagram of Circuit (Fig. 3-27)

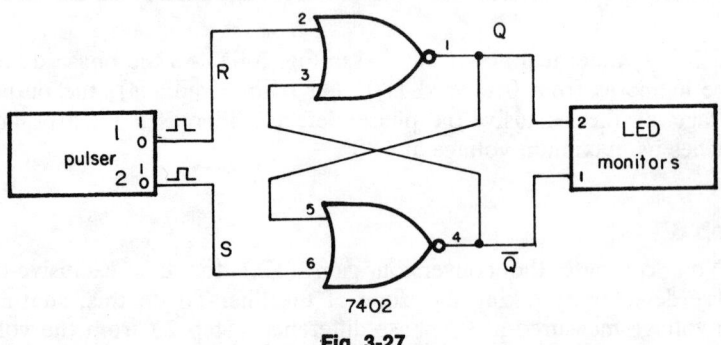

Fig. 3-27.

Step 1

Wire the circuit shown in the schematic diagram. Don't forget the power-supply connections to the 7402 integrated circuit!

Step 2

Apply power to the breadboard. Which LED is lit?

LED No. 2 should be lit indicating that output Q of the flip-flop is *high* or logic 1.

Step 3

Now press and release the No. 1 pulser so that the *reset* or *R* input goes from logic 0 to logic 1 and back to logic 0. What happens to the two LED monitors?

LED No. 2 is now unlit while LED No. 1 is lit, indicating that output Q is at logic 0 and output $\overline{Q}$ is at logic 1.

Step 4

Press and release the No. 1 pulser several times. Does anything happen?

You should observe that *nothing* happens! This is one basic characteristic of the R-S flip-flop:

If the reset or R input is at logic 1 and the set or S input is at logic 0, the output Q (LED No. 2) goes to or stays at logic 0, while output $\overline{Q}$ (LED No. 1) goes to or stays at logic 1.

Step 5

Now press and release pulser No. 2 so that the *set* or *S* input goes from logic 0 to logic 1 and back to logic 0. What happens?

LED No. 2 is now lit while LED No. 1 is unlit, indicating that output Q is at logic 1 and output $\overline{Q}$ is at logic 0.

Step 6

Press and release pulser No. 2 several times. Does anything happen?

Nothing happens! This is another basic characteristic of the R-S flip-flop:

If the set or S input is at logic 1 while the reset or R input is at logic 0, output Q (LED No. 2) goes to or stays at logic 1, while output $\overline{Q}$ (LED No. 1) goes to or stays at logic 0.

If both inputs are at any time simultaneously at logic 1, the flip-flop is said to be in a *disallowed condition*. Both outputs are then simultaneously at logic 0. However, the last input to go to logic 0 determines the final state. *This situation is to be avoided!*

EXPERIMENT NO. 5

Purpose

The purpose of this experiment is to demonstrate the dynamic characteristics of an edge-triggered detector, using a 7474 D-type TTL flip-flop.

Pin Configuration of Integrated-Circuit Chips (Fig. 3-28)

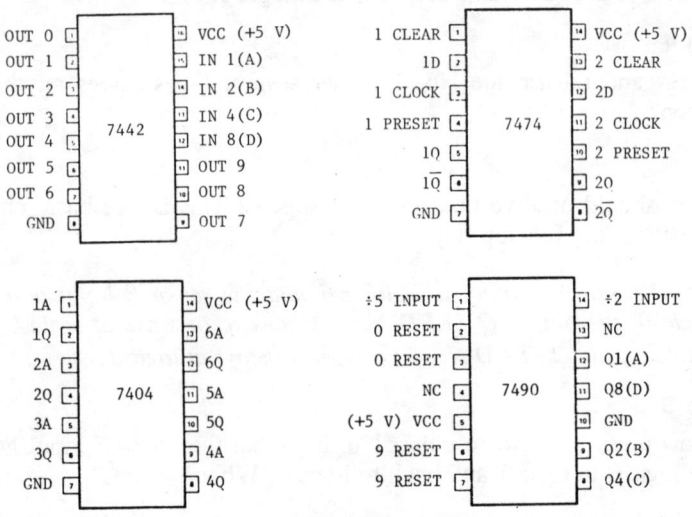

```
                  7442                                7474
OUT 0  [1]   [16] VCC (+5 V)        1 CLEAR  [1]   [14] VCC (+5 V)
OUT 1  [2]   [15] IN 1(A)                1D  [2]   [13] 2 CLEAR
OUT 2  [3]   [14] IN 2(B)         1 CLOCK   [3]   [12] 2D
OUT 3  [4]   [13] IN 4(C)         1 PRESET  [4]   [11] 2 CLOCK
OUT 4  [5]   [12] IN 8(D)               1Q  [5]   [10] 2 PRESET
OUT 5  [6]   [11] OUT 9                  1Q̄  [6]   [9]  2Q
OUT 6  [7]   [10] OUT 8                 GND  [7]   [8]  2Q̄
GND    [8]   [9]  OUT 7
```

```
                  7404                                7490
1A   [1]   [14] VCC (+5 V)       ÷5 INPUT  [1]   [14] ÷2 INPUT
1Q   [2]   [13] 6A                0 RESET  [2]   [13] NC
2A   [3]   [12] 6Q                0 RESET  [3]   [12] Q1(A)
2Q   [4]   [11] 5A                     NC  [4]   [11] Q8(D)
3A   [5]   [10] 5Q             (+5 V) VCC  [5]   [10] GND
3Q   [6]   [9]  4A                9 RESET  [6]   [9]  Q2(B)
GND  [7]   [8]  4Q                9 RESET  [7]   [8]  Q4(C)
```

Fig. 3-28.

Schematic Diagram of Circuit (Fig. 3-29)

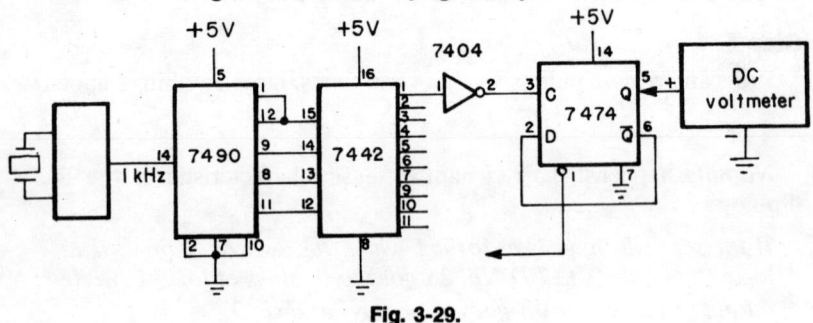

Fig. 3-29.

Step 1

Wire the circuit shown in the schematic diagram. Then apply power to the breadboard.

Step 2

Now connect the phase detector's *reset* input (pin 1 of the 7474 flip-flop) to pin 1 of the 7442 decoder chip. The portion composed of the 7490 decade counter and the 7442 decoder is a simple phase generator with fixed increments of 36°. The 7404 inverter is used to provide the proper *set* input pulse signal.

Step 3

Starting with pin 1 of the 7442 decoder, measure the dc output voltage of the phase detector at pin 5 of the 7474 chip, completing Table 3-8.

Table 3-8.

Reset Input (7442 Output Pin)	$\Delta\phi$	Output Voltage
1	0°	
2	36°	
3	72°	
4	108°	
5	144°	
6	180°	
7	216°	
9	252°	
10	288°	
11	324°	

Step 4

On the blank graph provided (Fig. 3-30), plot the results you obtained in Step 3. What striking difference do you notice about the characteristic curve of the edge-triggered phase detector as compared with the exclusive-OR phase detector?

You should have plotted a straight line from 0° to 324° so that the linear operating range of the edge-triggered phase detector is twice that of the exclusive-OR type.

Step 5

From your data, compute the conversion gain (K_ϕ) for this edge-triggered phase detector. Your results should be approximately as follows:

$$K_\phi = \frac{\Delta V_o}{\Delta \phi}$$

$$= \frac{3.44 \text{ volts} - 0.08 \text{ volt}}{(324° - 0°)(1 \text{ rad}/57.3°)}$$

$$= 0.594 \text{ V/rad}$$

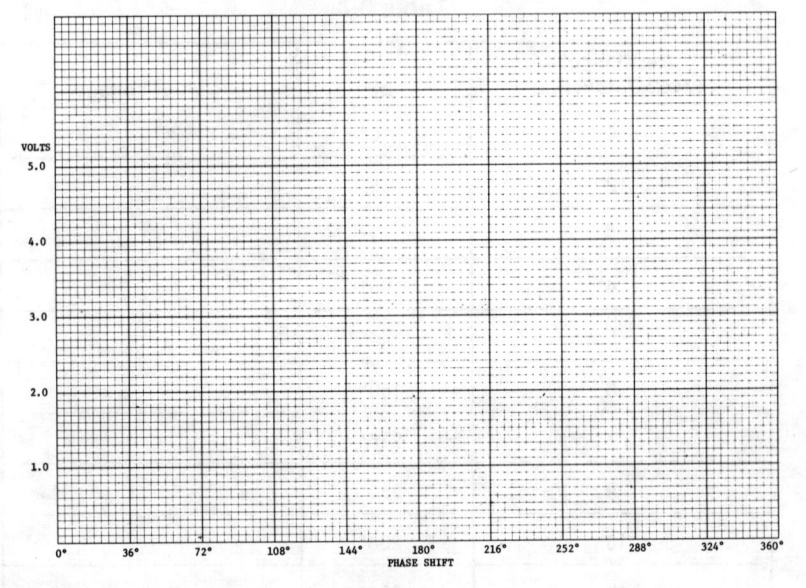

Fig. 3-30.

which is approximately one-half that of the exclusive-OR phase detector.

EXPERIMENT NO. 6

Purpose

The purpose of this experiment is to demonstrate the operation of the Motorola MC4044 integrated-circuit phase detector.

Pin Configuration of Integrated-Circuit Chips (Fig. 3-31)

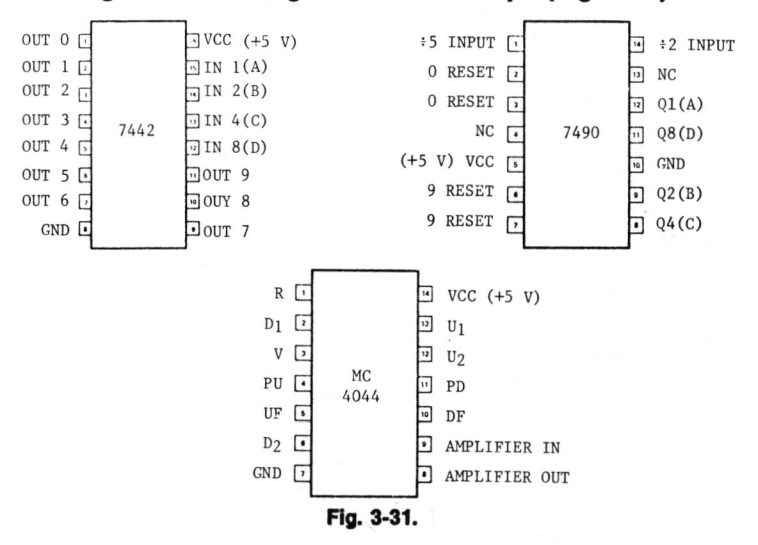

```
OUT 0  [1]            [16] VCC (+5 V)         ÷5 INPUT [1]          [14] ÷2 INPUT
OUT 1  [2]            [15] IN 1(A)            0 RESET  [2]          [13] NC
OUT 2  [3]            [14] IN 2(B)            0 RESET  [3]          [12] Q1(A)
OUT 3  [4]   7442     [13] IN 4(C)               NC    [4]  7490    [11] Q8(D)
OUT 4  [5]            [12] IN 8(D)            (+5 V) VCC [5]        [10] GND
OUT 5  [6]            [11] OUT 9              9 RESET  [6]          [9] Q2(B)
OUT 6  [7]            [10] OUY 8              9 RESET  [7]          [8] Q4(C)
GND    [8]            [9] OUT 7
```

```
         R   [1]              [14] VCC (+5 V)
         D1  [2]              [13] U1
         V   [3]              [12] U2
         PU  [4]   MC         [11] PD
         UF  [5]   4044       [10] DF
         D2  [6]              [9] AMPLIFIER IN
         GND [7]              [8] AMPLIFIER OUT
```

Fig. 3-31.

Schematic Diagram of Circuit (Fig. 3-32)

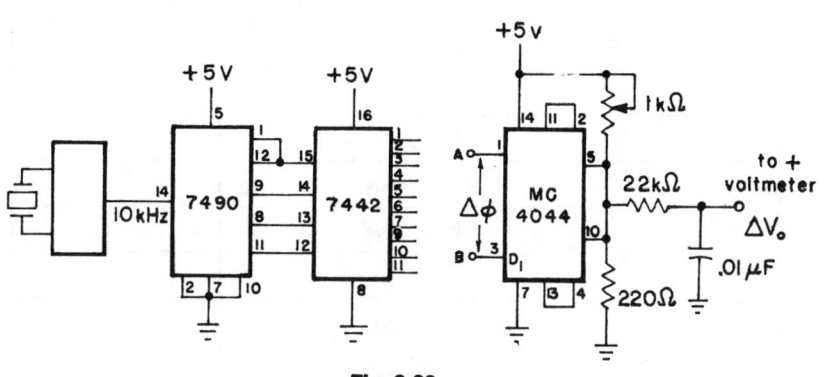

Fig. 3-32.

Step 1

Wire the circuit shown in the schematic diagram. Initially connect pins 1 and 3 of the MC4044 phase detector to pin 1 of the 7442 de-

49

coder ($\Delta\phi = 0°$). Apply power to the breadboard and then connect the 10-kHz, TTL-level square-wave signal source to the input of the 7490 decade counter. Then adjust the 1-kΩ potentiometer so that the output voltage (V_o) equals 1.50 volts.

Step 2

Connect the two inputs, A and B, of the MC4044 as directed in Table 3-9, and record your results.

Table 3-9.

MC4044 Inputs A	B (Connect to 7442 Pin No.:)	$\phi\Delta$	V_o
1	1	0° (in phase)	1.50 (adjusted)
2	1	−324° (B lags A)	
3	1	−288°	”
4	1	−252°	”
5	1	−216°	”
6	1	−180°	”
7	1	−144°	”
9	1	−108°	”
10	1	−72°	”
11	1	−36°	”
1	2	+324° (B leads A)	
1	3	+288°	”
1	4	+252°	”
1	5	+216°	”
1	6	+180°	”
1	7	+144°	”
1	9	+108°	”
1	10	+72°	”
1	11	+36°	”

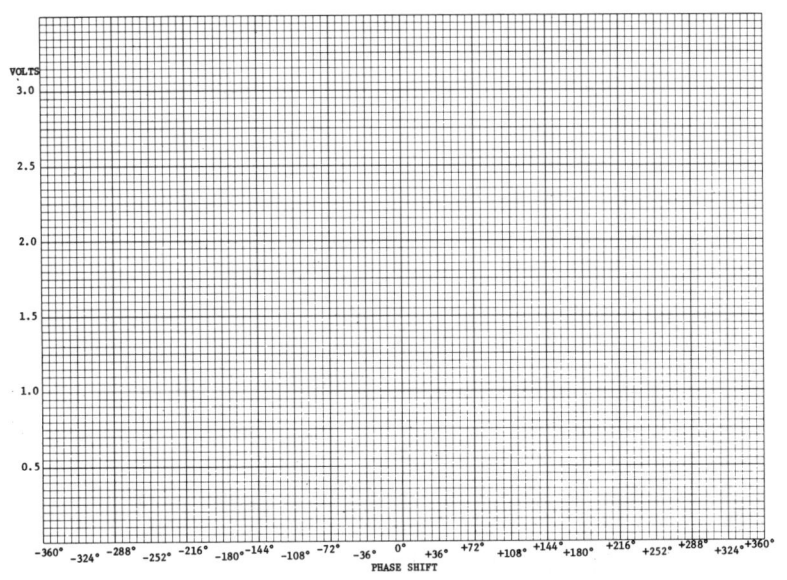

Fig. 3-33.

Step 3

Now plot your results on the blank graph (Fig. 3-33) provided for this purpose. If you have done everything correctly, you should find that the linear operating range of the MC4044 phase detector is twice as great as the range of the edge-triggered detector (i.e., 4π versus 2π radians).

Step 4

Based on your measurements in Step 2, determine the conversion gain for the MC4044. Your results should be approximately as follows:

$$K_\phi = \frac{\Delta V_o}{\Delta \phi}$$

so that,

$$\Delta \phi = \frac{648°}{57.3°/\text{rad}} = 11.31 \text{ radians}$$

$$\Delta V_o = 2.10 - 0.96 = 1.14 \text{ volts}$$

$$K_\phi = \frac{1.14 \text{ volts}}{11.31 \text{ radians}} = 0.10 \text{ V/rad}$$

as compared with a typical value of 0.12 V/rad (Motorola).

CHAPTER 4

The Voltage-Controlled Oscillator

INTRODUCTION

The voltage-controlled oscillator, or vco, is the second integral building block of the phase-locked loop, whose output frequency is that of the vco. This chapter discusses the operation and design of several circuits built around the MC4024 and MC1648 integrated circuits.

OBJECTIVES

At the completion of this chapter, you will be able to do the following:

- Describe the basic function of the phase-locked-loop vco.
- Define vco conversion gain.
- Design vco circuits using the MC4024 and MC1648 integrated circuits.
- Explain the function of a voltage-variable capacitor.

VCO BASICS

The voltage-controlled oscillator, or vco for short, is a network whose output frequency is directly proportional to its input control voltage (Fig. 4-1).

```
input voltage  ──→  │  VCO   │  ──→ output frequency
     V_f            │ (K_o)  │           ω_o
```

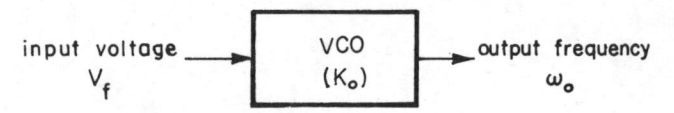

Fig. 4-1. Block diagram of the basic voltage-controlled oscillator.

The vco can also be termed a voltage-to-frequency converter, so that mathematically,

$$\omega_o = K_o V_f \qquad \text{(Eq. 4-1)}$$

where,
 ω_o is the vco output frequency (rad/s),
 V_f is the vco input control voltage from the loop filter,
 K_o is the vco conversion gain (rad/s/V).

The vco conversion gain (K_o) is the proportionality constant that converts the input control voltage to frequency.

In almost all the literature on phase-locked loops, the Greek letter ω (omega) is used to represent frequency, in units of radians/second. The *radian* frequency (ω) is related to the frequency (f) expressed in Hz (hertz), by the factor 2π, so that

$$\omega = 2\pi f \qquad \text{(Eq. 4-2)}$$

For example, the 60-Hz power line frequency is equal to the radian frequency of 2π times 60 Hz, or 377 rad/s.

During phase-lock, the output frequency of the vco will be exactly equal to the input frequency of the loop, except for a constant finite phase difference. Since the radian frequency is the time derivative of phase,

$$\omega = \frac{d(\Delta\phi)}{dt} \quad \text{(rad/s)} \qquad \text{(Eq. 4-3)}$$

the phase difference between the vco output frequency and the loop input frequency is really proportional to the integral of the input control voltage, which is the average or dc voltage from the phase detector and the loop filter. Any ac signal superimposed on the dc control voltage will, in turn, vary the vco frequency.

VCO CIRCUITS

Although there are a myriad of circuits possible, we will consider two of the more popular vco integrated circuits that are used in phase-locked-loop systems.

MC4024 Voltage-Controlled Multivibrator (VCM)

The MC4024 integrated circuit is a 14-pin DIP device (shown in Fig. 4-2) made by Motorola. It contains two independent voltage-controlled multivibrators (square-wave generators) with output buffers. Each vcm contains approximately 17 transistors, 23 resistors, and 5 diodes.

From the pin diagram of Fig. 4-2, it should be noted that there are three $+V_{CC}$ (+5 volts) and three ground connections provided. Each vcm section has separate $+V_{CC}$ and ground connections, and the corresponding output buffer has a common $+V_{CC}$ and ground pin. This method provides isolation between the two vcm sections while minimizing the effect of output buffer transients on the vcm in critical applications.

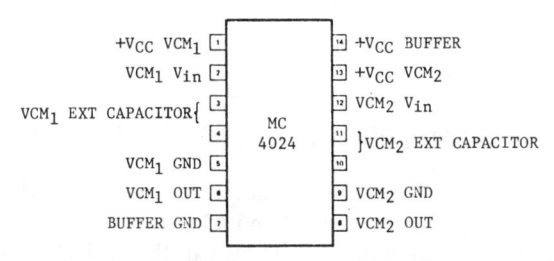

Fig. 4-2. Pin diagram of the MC4024 voltage-controlled multivibrator.

The operating frequency range is controlled by the value of an external capacitor connected between pins 3 and 4 (or 10 and 11), as shown in the basic circuit of Fig. 4-3.

For values of the external capacitor greater than 100 pF, the approximate frequency can be determined from

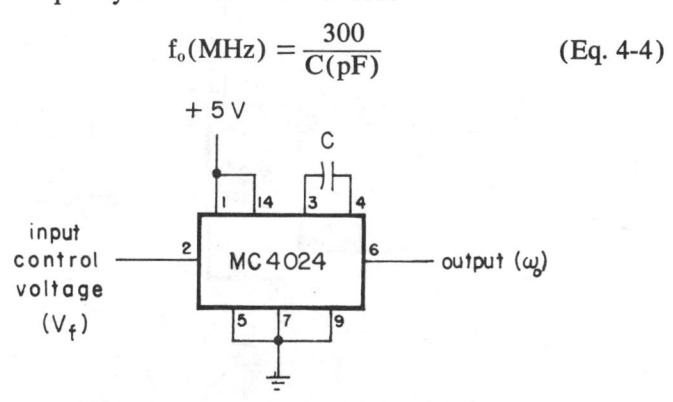

$$f_o(\text{MHz}) = \frac{300}{C(\text{pF})} \qquad \text{(Eq. 4-4)}$$

Fig. 4-3. Circuit diagram of the MC4024.

55

The MC4024 has a maximum output frequency of 25 MHz, and under ideal conditions, a tuning range of 3.5 to 1. For complete design information, consult the MC4024 data sheet given in Appendix B.

MC1648 Voltage-Controlled Oscillator

The MC1648 integrated circuit, also made by Motorola, is an emitter-coupled (ECL) device housed in a 14-pin package (Fig. 4-4). It contains the equivalent of 11 transistors, 13 resistors, and 2 diodes forming an oscillator with an output buffer.

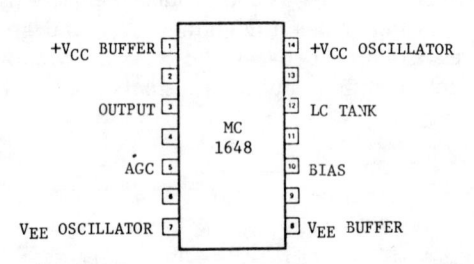

Fig. 4-4. Pin diagram of the MC1648 voltage-controlled oscillator.

Normally powered by a +5-volt supply (V_{EE} = ground), the MC-1648 requires an external parallel inductor-capacitor, or LC "tank" network to produce oscillation, as shown in the typical operating circuit of Fig. 4-5. For this circuit, the square-wave oscillator frequency is determined from

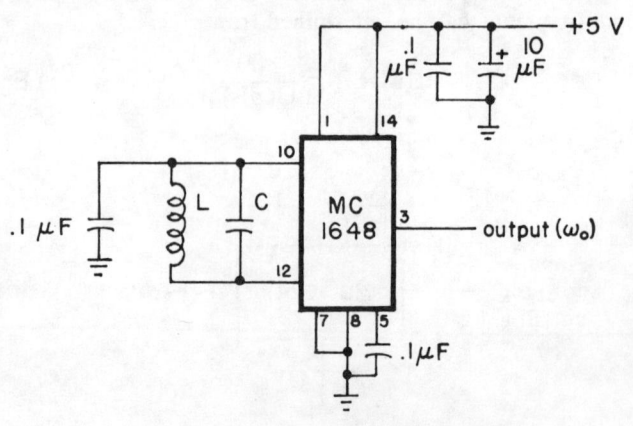

$$\omega_o = \frac{1}{(LC_T)^{1/2}} \quad \text{(rad/s)} \qquad \text{(Eq. 4-5a)}$$

Fig. 4-5. Circuit diagram of the MC1648.

or,

$$f_o = \frac{1}{2\pi(LC_T)^{1/2}} \text{ (Hz)} \qquad \text{(Eq. 4-5b)}$$

where C_T is the sum of the capacitance of the oscillator tank circuit (C) and the input capacitance (C_{in}) of the MC1648, which is typically 6 pF. With suitable values for the external tank circuit, it is possible to achieve output frequencies up to about 225 MHz.

When the MC1648 is used as a vco in phase-locked-loop systems, a voltage-controlled diode, called a *varactor diode,* is generally used as part of the tank circuit to provide a voltage variable input for the vco, as shown in the circuit of Fig. 4-6.

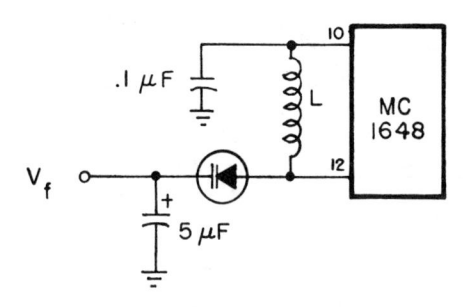

Fig. 4-6. MC1648 varactor-diode circuit.

The tuning range of the vco is then determined from

$$\frac{f_{max}}{f_{min}} = \left(\frac{C_{D(max)} + C_{in}}{C_{D(min)} + C_{in}}\right)^{1/2} \qquad \text{(Eq. 4-6)}$$

where,

$$f_{min} = \frac{1}{2\pi[L(C_{D(max)} + C_{in})]^{1/2}}, \qquad \text{(Eq. 4-7)}$$

C_D = varactor capacitance of input bias voltage,
C_{in} = input capacitance of MC1648 (typically 6 pF).

In the next section we shall describe the use and selection of varactor diodes.

THE VARACTOR

As briefly pointed out in the previous section, a varactor diode is generally used to vary the capacitance of an oscillator LC tank network. Basically a varactor is a voltage-variable capacitor, based on semiconductor phenomena which can be used for electronic tuning applications. In operation, the varactor operates principally in the region between forward conduction and reverse breakdown, which is

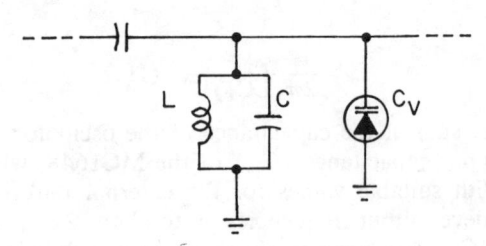

Fig. 4-7. Parallel varactor tuning.

the region that a conventional diode is considered to be cut off. Consequently, the varactor operates neither as a rectifier (forward conduction) nor as a zener diode (reverse breakdown).

The majority of the varactor resonant circuits take the form of either Fig. 4-7 (parallel tuned) or Fig. 4-8 (series tuned). In either case, the resonant or oscillation frequency of the tank circuit is given by Equation 4-5, where capacitance C_T is now the sum of the capacitances of the tank circuit, the MC1648 input, and the varactor.

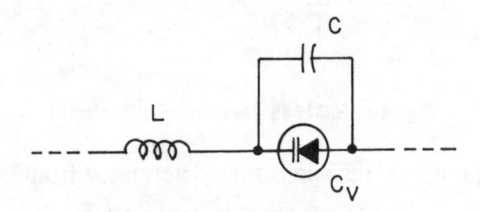

Fig. 4-8. Series varactor tuning.

To choose the proper varactor, three parameters are considered:

1. C_T—Nominal capacitance, usually at a specified voltage.
2. Capacitance Ratio (CR)—The ratio of the capacitance at two separate applied voltages, so that

$$CR = \frac{C_{V(min)}}{C_{V(max)}} = \left(\frac{V_{max}}{V_{min}}\right)^\rho \qquad \text{(Eq. 4-8)}$$

where,
   ρ = capacitance exponent (typically 0.5, but can vary from 0.3 to 2.0).

3. Frequency Ratio (FR)—Equal to the square root of the capacitance ratio,

$$FR = (CR)^{1/2} \qquad \text{(Eq. 4-9)}$$

Example

As an illustration, the MV2107 varactor (Motorola) has a nominal capacitance of 22 pF (±10%) at 4 volts. The capacitance ratio is 2.5 to 1 over a voltage range of 2 to 30 volts, resulting in a frequency ratio of 1.58 to 1. From Equation 4-8 the capacitance exponent (ρ) can be determined for the MV2107 varactor,

$$CR = \left(\frac{V_{max}}{V_{min}}\right)^\rho$$

$$2.5 = \left(\frac{30}{2}\right)^\rho$$

$$= (15)^\rho$$

$$\log(2.5) = \rho \log(15)$$

so that,

$$\rho = 0.3384$$

Now, using Equation 4-8 again and the nominal value of 22 pF at 4 volts,

$$\frac{C_{V(min)}}{C_{V(max)}} = \left(\frac{V_{max}}{V_{min}}\right)^\rho$$

$$\frac{22 \text{ pF}}{C_{V(max)}} = \left(\frac{30 \text{ V}}{4 \text{ V}}\right)^{0.3384}$$

$$= 1.98$$

Solving for $C_{V(max)}$ we obtain,

$$C_{V(max)} = 11.1 \text{ pF } (@ \ 30 \text{ volts})$$

Since the capacitance ratio is 2.5, then once more from Equation 4-8 we solve for $C_{V(min)}$,

$$CR = \frac{C_{V(min)}}{C_{V(max)}}$$

$$2.5 = \frac{C_{V(min)}}{11.1 \text{ pF}}$$

or,

$$C_{V(min)} = 27.8 \text{ pF } (@ \ 2 \text{ volts})$$

As an extension of this example, we can see how this capacitance variation affects the resonant frequency of the tank circuit of a 1-μH inductor and the MV2107 varactor. At 2 volts, the resonant frequency from Equation 4-7 is

$$f_{o(min)} = \frac{1}{2\pi[(1\ \mu H)(27.8\ pF)]^{1/2}}$$
$$= 30.2\ \text{MHz}\ (@\ 2\ \text{volts})$$

Using Equation 4-7 again, but to calculate $f_{o(max)}$ with $C = 11.1$ pF, the frequency is

$$f_{o(max)} = 47.8\ \text{MHz}\ (@\ 30\ \text{volts})$$

so that the frequency ratio $f_{o(max)}/f_{o(min)}$ is

$$\frac{f_{o(max)}}{f_{o(min)}} = \frac{47.8\ \text{MHz}}{30.2\ \text{MHz}}$$
$$= 1.58$$

which is exactly the frequency ratio determined earlier.

OTHER INTEGRATED CIRCUITS

Other integrated circuits that are able to function as a vco are available. However, their maximum frequency, usually determined by an external resistor and capacitor, is generally limited to about 1 MHz, and they are not suited for the frequency ranges required by phase-locked-loop synthesizers (Chapter 6). Examples of the devices that fall into this group are the XR-2206 function generator (Exar), 8038 waveform generator/vco (Intersil), and the 566 vco (Signetics and National Semiconductor). On the other hand, there are monolithic integrated circuits that contain a vco within a phase-locked-loop system on a single chip. These devices are discussed in Chapter 7.

AN INTRODUCTION TO THE EXPERIMENT

The sole experiment in this chapter examines the operation of the MC4024 voltage-controlled multivibrator, whose output frequency is a function of the input control voltage for a given external frequency-determining capacitor.

EXPERIMENT NO. 1

Purpose

The purpose of this experiment is to demonstrate the operation of a voltage-controlled oscillator, using an MC4024 integrated circuit.

Pin Configuration of Integrated-Circuit Chip (Fig. 4-9)

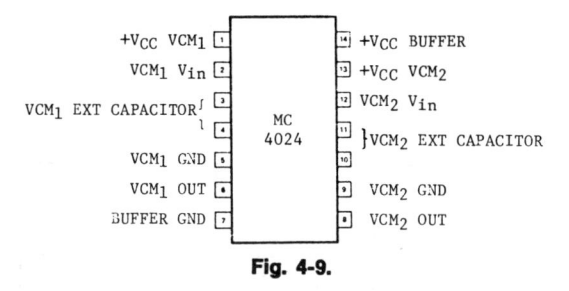

```
+VCC VCM1   [1]        [14] +VCC BUFFER
VCM1 Vin    [2]        [13] +VCC VCM2
VCM1 EXT CAPACITOR {[3] MC  [12] VCM2 Vin
                   {[4] 4024 [11]} VCM2 EXT CAPACITOR
VCM1 GND    [5]        [10]
VCM1 OUT    [6]        [9]  VCM2 GND
BUFFER GND  [7]        [8]  VCM2 OUT
```

Fig. 4-9.

Schematic Diagram of Circuit (Fig. 4-10)

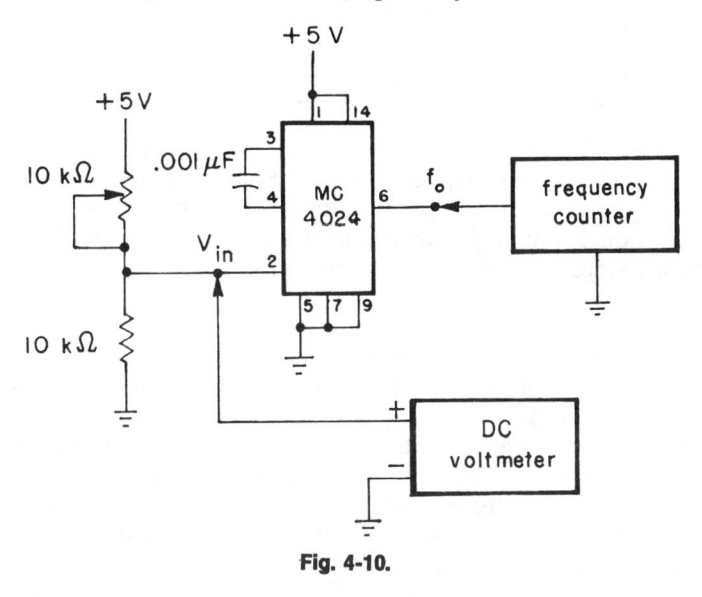

Fig. 4-10.

Step 1

Wire the circuit shown in the schematic diagram and apply power to the breadboard. With the voltmeter connected to pin 2 of the MC-4024 integrated circuit, adjust the 10-kΩ potentiometer until the voltmeter reads +2.50 volts, which is the input control voltage of the vco.

Step 2

Now vary the input control voltage by varying the potentiometer according to Table 4-1, recording the resultant output frequency of the vco.

Table 4-1.

V_{in}	f_o
2.50 V	
2.75 V	
3.00 V	
3.25 V	
3.50 V	
3.75 V	
4.00 V	
4.25 V	
4.50 V	

You should have observed that, as the input control voltage is increased, the output frequency of the vco also increased. The exact frequency at any given input control voltage isn't really important for this experiment. This experiment is merely showing that the input control voltage controls the vco output frequency.

Step 3

Disconnect the power to the breadboard and change the capacitor connected between pins 3 and 4 to 100 pF. Apply power to the breadboard and set the input control voltage at 2.50 volts. You should have measured an output frequency that is approximately 10 times higher than in Step 2. For the MC4024 device, as with other types of voltage-controlled oscillators, the value of an external frequency-determining capacitor also affects the vco frequency as well as the input control voltage.

CHAPTER 5

The Loop Filter and Loop Response

INTRODUCTION

So far, we have discussed the phase detector and the voltage-controlled oscillator. However, nothing has been said about what determines the overall loop response. It is the job of the loop filter to control the lock, capture, bandwidth, and transient response of the loop. In this chapter, several types of low-pass filter networks are discussed to accomplish these functions.

OBJECTIVES

At the completion of this chapter, you will be able to do the following:

- Describe the effect of the loop response as the damping factor is varied.
- Define the following terms:
    natural frequency
    damped natural frequency
    bandwidth
    settling time
    overshoot
    lock range
    capture
- Describe the lock and capture process by means of a voltage-to-frequency diagram.

- Describe several low-pass filter networks used in phase-locked-loop systems.
- Experimentally determine the parameters of a simple second-order system.

FUNCTION OF THE LOOP FILTER

The inclusion of a low-pass filter network in the phase-locked loop has two major functions. First, it removes any noise and high-frequency components from the output voltage of the phase detector, thus giving an average (dc) voltage. Second, it is the primary building block that determines the dynamic performance of the loop, which includes the following factors:

- Capture and lock ranges.
- Bandwidth.
- Transient response.

The loop filter may either be passive or active. Although a detailed mathematical discussion is needed for an appreciation of why the filter does what it is supposed to do, it is nevertheless beyond the scope of this book. However, the derivations for most of the design equations presented in this chatper are given in Appendix A.

The complete phase-locked-loop system exhibits the characteristics of a *second-order system* (analogous to a swinging pendulum or vibrating string). Before we can describe any of the possible filter circuits, we must first have a general idea of what factors affect the loop response, since these parameters are in turn determined by the filter design. As a function of frequency, the response of a second-order system has the form:

$$\frac{V_{out}}{V_{in}} \text{ (dB)} = -20 \log [\omega^4 + 2\omega^2(2\zeta^2 - 1) + 1]^{1/2}$$

(Eq. 5-1)

where,

V_{in} = loop input voltage,
V_{out} = loop output voltage,
ζ = damping factor (dimensionless),
ω = ratio of the input frequency (ω_1) to the undamped natural frequency (ω_n).

By plotting Equation 5-1, as shown in Fig. 5-1, the single parameter that governs the overall shape of the response-versus-frequency curve of a second-order system is the *damping factor* (ζ), which has been called the *damping ratio* by some writers. For a given value of damping, the frequency at which the response is a maximum is the *undamped natural frequency,* or *natural frequency* for short. For a lesser

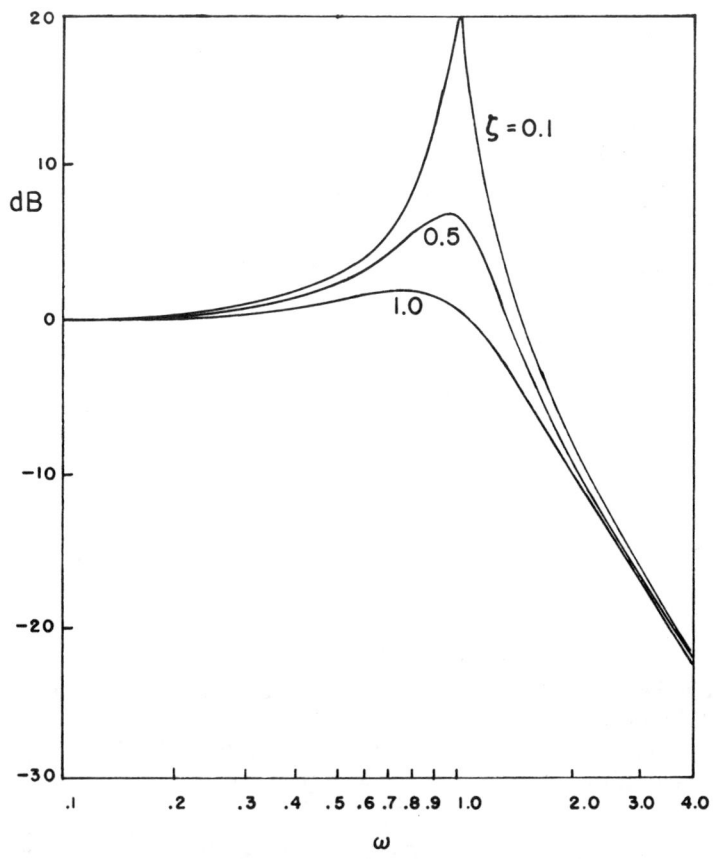

Fig. 5-1. Effect of damping on the frequency response of a second-order system.

amount of damping, we have a greater amount of peaking at the natural frequency. The frequency at which the response is 3 dB less than the maximum response is called the *bandwidth* of the system. If the damping factor were allowed to be zero, we would then have a sinusoidal oscillator (see Appendix A). The parameters such as the damping factor and the undamped natural frequency are primarily controlled by the loop filter. That is, depending on the filter design, we are able to control the loop response.

LOW-PASS FILTER CIRCUITS

In phase-locked-loop systems there are a number of popular low-pass filter circuits. In this section three such types will be discussed.

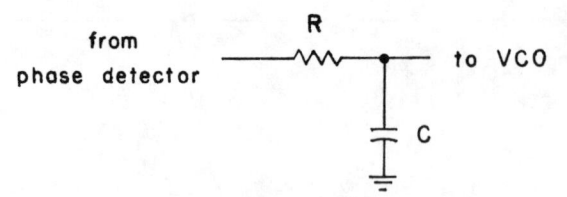

Fig. 5-2. First-order RC low-pass filter.

As shown in Fig. 5-2, we have a simple first-order RC network which is placed between the phase detector and the vco. The *cutoff frequency* (ω_{LPF}) for this filter is given by

$$\omega_{LPF} = 1/RC \quad (\text{rad/s}) \qquad (\text{Eq. 5-2})$$

Without any derivation, the loop *natural frequency* can be expressed in terms of the filter cutoff frequency, so that

$$\omega_n = (K_\phi K_o \omega_{LPF})^{1/2} \qquad (\text{Eq. 5-3})$$

In addition, the damping factor can be written as

$$\zeta = \frac{1}{2}\left(\frac{\omega_{LPF}}{K_\phi K_o}\right)^{1/2} \qquad (\text{Eq. 5-4})$$

How the natural frequency and damping factor of the loop are chosen will be explained in the next section.

Another passive filter network is the *lag-lead* circuit of Fig. 5-3. The cutoff frequency for this filter is given by

$$\omega_{LPF} = \frac{1}{(R_1 + R_2)C} \quad (\text{rad/s}) \qquad (\text{Eq. 5-5})$$

The natural frequency is then written as

$$\omega_n = (K_\phi K_o \omega_{LPF})^{1/2} \quad (\text{rad/s}) \qquad (\text{Eq. 5-6})$$

and the damping factor is given by

$$\zeta = \frac{\omega_n}{2}\left[R_2 C + \left(\frac{1}{K_\phi K_o}\right)\right] \qquad (\text{Eq. 5-7})$$

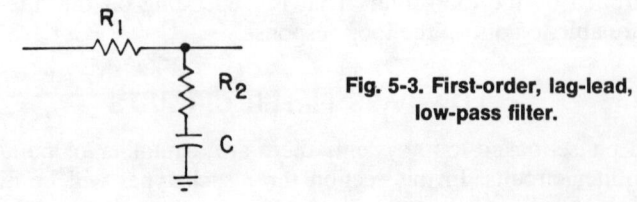

Fig. 5-3. First-order, lag-lead, low-pass filter.

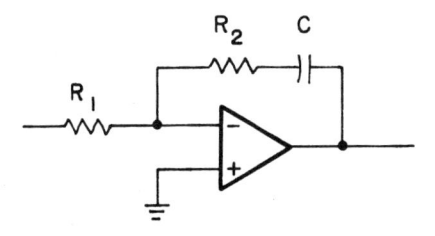

Fig. 5-4. Active low-pass filter.

The passive lag-lead network can be used with an operational amplifier to form an active filter circuit, as shown in Fig. 5-4. The cutoff frequency is written as

$$\omega_{LPF} = \frac{1}{R_1 C} \quad (rad/s) \quad \text{(Eq. 5-8)}$$

while the loop natural frequency and damping factor are found from

$$\omega_n = (K_\phi K_o \omega_{LPF})^{1/2} \quad (rad/s) \quad \text{(Eq. 5-9)}$$

and,

$$\zeta = \left(\frac{R_2 C}{2}\right)\omega_n \quad \text{(Eq. 5-10)}$$

In Equations 5-3, 5-6, and 5-9, the natural frequency of the loop depends entirely on the product $K_\phi K_o$ (often referred to as the *dc loop gain*) and the filter cutoff frequency.

THE TRANSIENT RESPONSE

When an underdamped second-order system ($\zeta < 1$) encounters a sudden change at its input, such as the phase-locked loop shifting from one frequency (f_1) to another (f_2), the output of the vco tries to follow this change but oscillates about the value of f_2 for a time and eventually settles out at the new frequency (i.e., steady state). This process is illustrated in Fig. 5-5. How fast this process is completed depends on the loop damping factor, which, in turn, is controlled by the loop filter. As shown in the graph of Fig. 5-6, when the simple RC filter of Fig. 5-2 is used, it takes longer for the oscillations to settle down to the steady-state value as the damping factor is decreased. To reasonably pick values for ζ and ω_n, one method is to design for a specified amount of *overshoot* within a given *settling time*.

- *Overshoot is the maximum difference between the transient and the steady-state value for a sudden change applied to the input of the phase-locked loop.*

Fig. 5-5. Transient response.

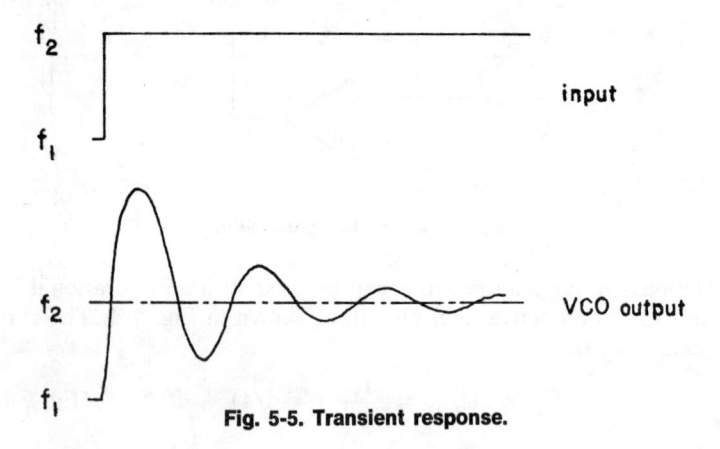

- *The settling time (t_s) is the time required for the transient response to reach and remain within a specified percentage of the steady-state value (e.g., 10%).*

As a general rule, the damping factor is chosen to be between 0.5 and 0.8. That leaves only the amount of overshoot and the settling time to decide upon, as illustrated in the following example:

Example

Using the loop filter of Fig. 5-2, suppose we want to design a phase-locked-loop system with a damping factor of 0.5 so that the output will be less than 10% of the steady-state value 10 ms after a change in frequency (or phase) is applied at the loop input.

From Fig. 5-6, we see that the response reaches and remains within 10% of the steady-state value at $\omega_n t = 4.5$ for a damping factor of 0.5. Since the settling time is 10 ms, the loop natural frequency is then found from

$$\omega_n t_s = 4.5$$

$$\omega_n = \frac{4.5}{t_s}$$

$$= \frac{4.5}{10 \text{ ms}}$$

$$= 450 \text{ rad/s } (71.6 \text{ Hz})$$

Then, by picking a value for the filter capacitor, and knowing the conversion gains of the vco and phase detector, we can determine the value of R from Equation 5-3.

An alternate approach is to design the filter based on its cutoff frequency. How far the cutoff frequency can be placed *below* the input frequency of the loop is not easily determined. As a starting point, the cutoff frequency is made 100 times smaller than the input frequency. The phase-locked-loop system is then tested by applying a sudden change in frequency at the loop input and observing the output

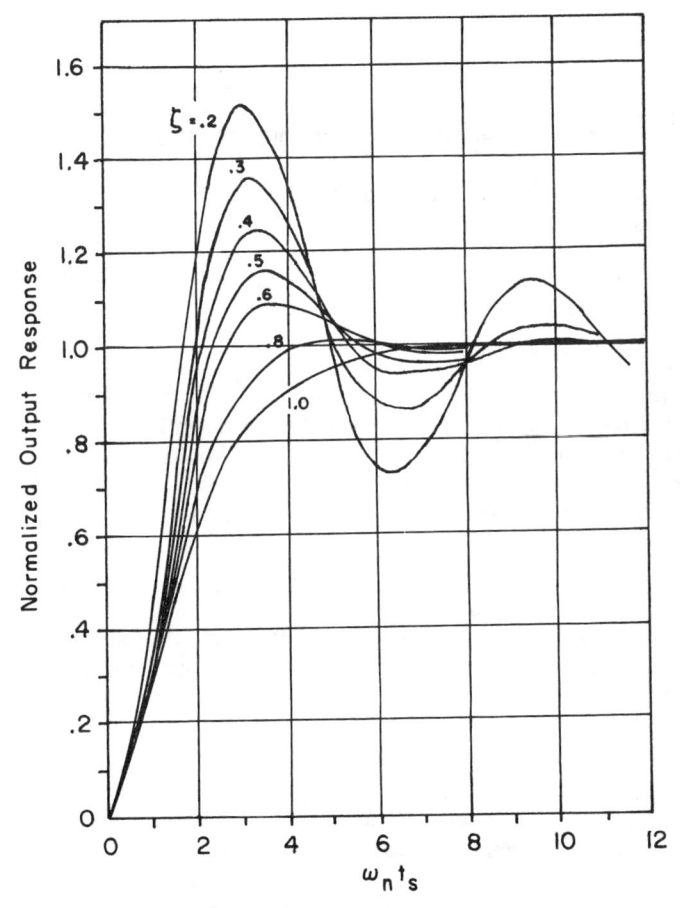

Fig. 5-6. Normalized transient response.

of the filter on the oscilloscope. The components of the filter are then adjusted to give the desired amount of overshoot and settling time.

From the transient response of the loop, the damping factor can be easily estimated simply by knowing the peak amplitude of two consecutive positive peaks which are exactly 1 cycle apart, as illus-

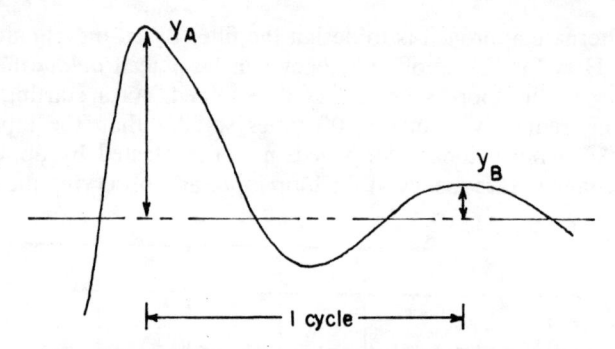

Fig. 5-7. Illustration for determining damping factor.

trated in Fig. 5-7. The damping factor is then determined from the relationship

$$\zeta = \frac{\gamma}{(1+\gamma^2)^{1/2}} \qquad \text{(Eq. 5-11)}$$

where,
$\gamma = (1/2\pi)\ln(y_A/y_B)$

If the loop filter of either Fig. 5-3 or 5-4 is placed between the phase detector and the vco, the resultant transient response is given by the curves of Fig. 5-8. In either case, we should notice that the response is periodic with a fixed frequency. This oscillatory frequency associated with this transient behavior is called the *damped natural frequency*, ω_d, so that

$$\omega_d = 2\pi/T \qquad \text{(Eq. 5-12)}$$

where T is the period of oscillation. However, the damped natural frequency depends on the damping factor and the natural frequency of the loop, so that

$$\omega_d = \omega_n(1-\zeta^2)^{1/2} \qquad \text{(Eq. 5-13)}$$

Consequently, the damped natural frequency of the transient response is always less than the loop natural frequency. How much of a difference, of course, depends on the damping factor.

LOCK AND CAPTURE

The *lock range* ($2\omega_L$) of the phase-locked loop is the frequency range over which the loop system will follow changes in the input frequency. Several writers use the terms *tracking range* and *hold-in range*. The hold-in range refers to how far the input frequency can deviate from the vco free-running frequency, ω_o, and is numerically one-half the lock, or tracking range.

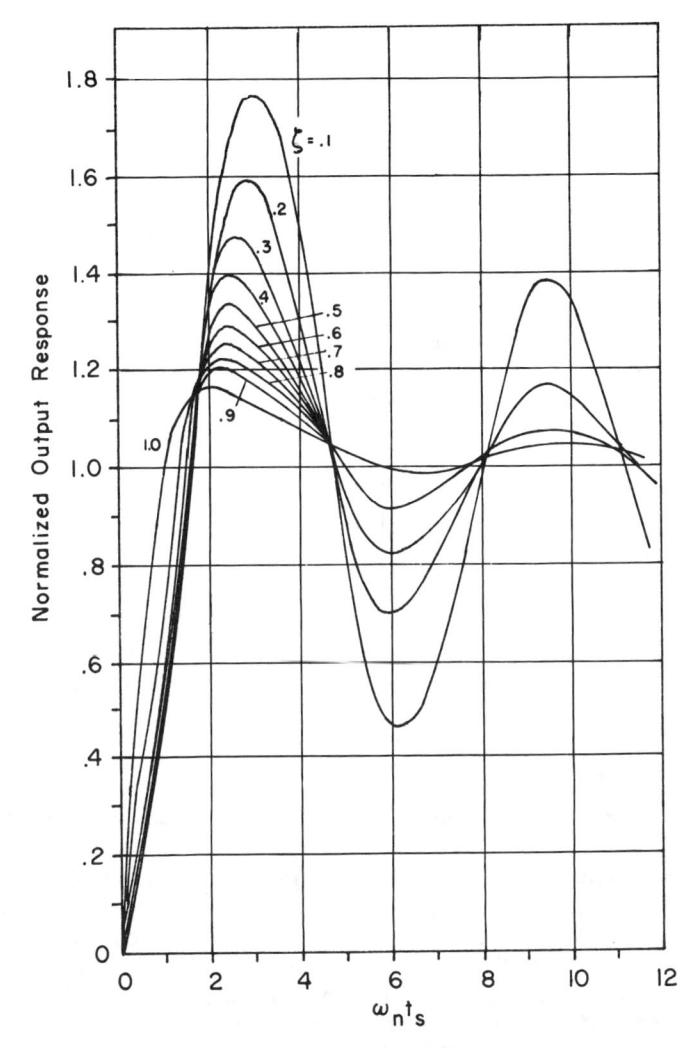

Fig. 5-8. Normalized transient response.

On the other hand, the range over which the phase-locked loop acquires phase lock is the *capture range* ($2\omega_C$). Several writers use the term *lock-in range,* which refers to how close an input frequency must be to the vco free-running frequency before the loop acquires phase-lock. The lock-in range is numerically one-half the capture range.

Fig. 5-9 shows the general frequency-to-voltage transfer characteristic of a phase-locked loop. In the top characteristic, the input

71

frequency (ω_i) is gradually increased so that the loop does not respond until ω_i equals ω_1, which is the lower edge of the capture range. The loop is then phase-locked onto the input frequency, causing the loop error voltage to go negative. As the input frequency is increased further, the error voltage increases linearly with a slope equal to the reciprocal of the vco conversion gain, or $1/K_o$ (V/rad/s). When the input frequency equals the vco free-running frequency, the error

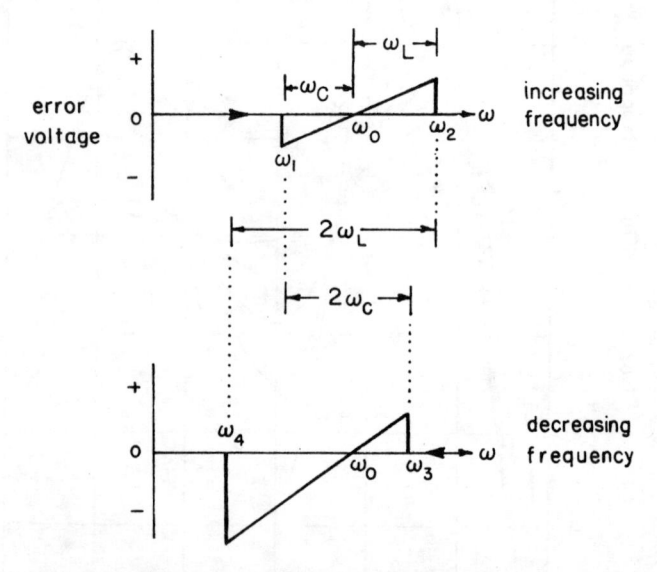

Fig. 5-9. Phase-locked-loop, frequency-to-voltage transfer characteristic.

voltage is zero. The loop continues to track the input until ω_2, the upper edge of the *lock range*. For input frequencies greater than ω_2, the loop is unlocked, the error voltage is zero, and the vco is at its free-running frequency. When the input frequency decreases, the process is repeated, except that now the error voltage goes positive at ω_3, the upper edge of the *capture range*.

In summary, we have the following relationships:

$$\text{lock range: } 2\omega_L = \omega_2 - \omega_4 \qquad \text{(Eq. 5-14)}$$

$$\text{hold-in range: } \omega_L = \omega_2 - \omega_o \qquad \text{(Eq. 5-15)}$$
$$= \omega_o - \omega_4$$

$$\text{capture range: } 2\omega_C = \omega_3 - \omega_1 \qquad \text{(Eq. 5-16)}$$

$$\text{lock-in range: } \omega_C = \omega_o - \omega_1 \qquad \text{(Eq. 5-17)}$$
$$= \omega_3 - \omega_o$$

In terms of loop parameters, the hold-in range is numerically equal to the dc loop gain (K), so that

$$\omega_L = K \quad \text{(Eq. 5-18)}$$
$$= K_\phi K_o \quad (\text{rad/s})$$

Note that from Equation 5-18, *the hold-in range does not depend on the parameters of the low-pass filter.* However, the filter does limit the maximum rate at which phase-lock can occur, since the voltage across the filter capacitor(s) cannot charge instantaneously.

The expression for the lock-in range, however, is quite involved. Nevertheless, we are able to give *approximate* expressions for the lock-in range based on the type of loop filter used. For the simple RC filter of Fig. 5-2, the lock-in range is given as

$$\omega_C \cong \left(\frac{\omega_L}{RC}\right)^{1/2} \quad (\text{rad/s}) \quad \text{(Eq. 5-19)}$$

For the passive lag-lead network of Fig. 5-3,

$$\omega_C \cong \omega_L \left(\frac{R_2}{R_1 + R_2}\right) \quad \text{(Eq. 5-20)}$$

and for the active filter of Fig. 5-4,

$$\omega_C \cong \omega_L \left(\frac{R_2}{R_1}\right) \quad \text{(Eq. 5-21)}$$

By making use of the equations for the loop damping factor and natural frequency, the lock-in range can be further approximated by

$$\omega_C \cong 2\zeta\omega_n \quad \text{(Eq. 5-22)}$$

AN INTRODUCTION TO THE EXPERIMENT

The sole experiment in this chapter uses an active filter to simulate an underdamped second-order, phase-locked-loop system. You will examine the transient response of this second-order system by applying a square-wave input voltage which represents the shifting from one input frequency to another. From the transient response, you will determine the damping factor, damped natural frequency, natural frequency, and bandwidth.

EXPERIMENT NO. 1

Purpose

The purpose of this experiment is to demonstrate the behavior of an underdamped second-order system represented by an active filter.

Pin Configuration of Integrated-Circuit Chip (Fig. 5-10)

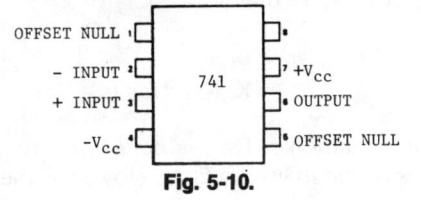

```
OFFSET NULL  1 ┌─────┐ 8
    - INPUT  2 │ 741 │ 7  +Vcc
    + INPUT  3 │     │ 6  OUTPUT
       -Vcc  4 └─────┘ 5  OFFSET NULL
```

Fig. 5-10.

Schematic Diagram of Circuit (Fig. 5-11)

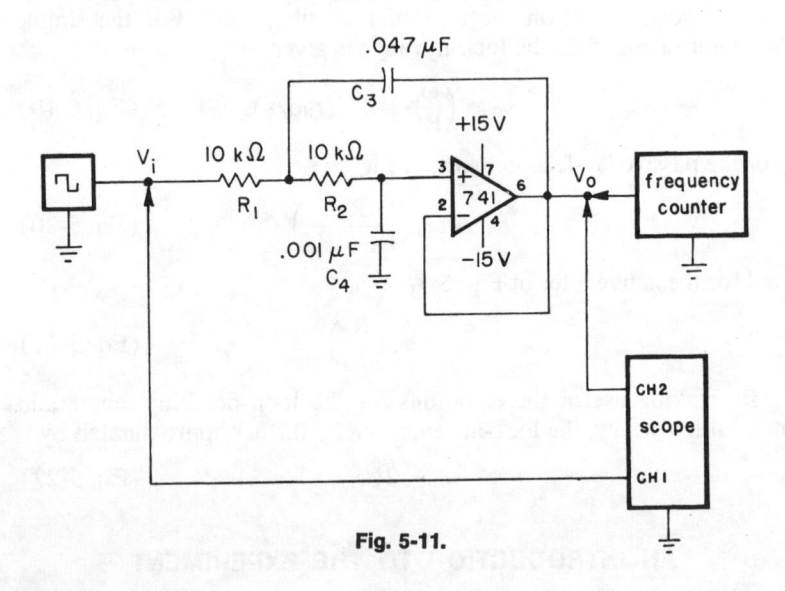

Fig. 5-11.

Design Basics

1. Natural frequency: $f_n = \dfrac{1}{2\pi(R_1 R_2 C_3 C_4)^{1/2}}$

2. Damping factor: $\zeta = \pi f_n C_4 (R_1 + R_2)$

3. Damped natural frequency: $f_d = f_n(1 - \zeta^2)^{1/2}$

4. 3-dB bandwidth: $f_{3\,dB} = f_n[1 + 2\zeta^2 + (2 - 4\zeta^2 + 4\zeta^4)^{1/2}]^{1/2}$

The equations for the natural frequency and the damping factor are taken from the book: *The Design of Active Filters, With Experiments,* by Howard M. Berlin. The equation for the 3-dB bandwidth is given in Appendix A of this book.

Step 1

Wire the circuit shown in the schematic diagram. If you do not have a dual-polarity power supply for the operational amplifier, you can easily make one from two 9-volt transistor radio batteries, as shown in Fig. 5-12.

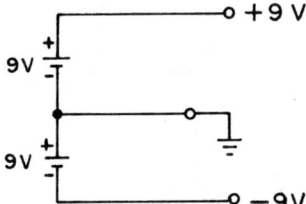

Fig. 5-12. Dual-polarity power supply.

Step 2

Set your oscilloscope for the following initial settings:

- Channel 1: 5 V/division
- Channel 2: 2 V/division
- Time base: 1 ms/division
- Triggering: Channel 1

Step 3

Apply power to the breadboard and adjust the output voltage of the function generator (square wave) at 5 volts peak-to-peak and the frequency at 150 Hz. If you have done everything correctly up to this point, the two traces on the oscilloscope screen should be similar to that shown in Fig. 5-13.

The input square wave of channel 1 is analogous to the shifting of the input frequency of the phase-locked loop. The bottom trace of channel 2 is the transient response of the underdamped second-order system, which is analogous to the loop error voltage. Note that it oscillates and eventually settles out at some steady-state value.

Step 4

Comparing your transient response with the general form shown in Fig. 5-7, determine the peak heights of *two successive* positive peaks, as measured from the steady-state value. Then from Equation 5-11, calculate the damping factor for this second-order system from your measurements and record your result:

$$\zeta = \underline{\qquad}$$

The peak heights of two successive peaks should be about 8 and 3 divisions, respectively, giving a damping factor of 0.145.

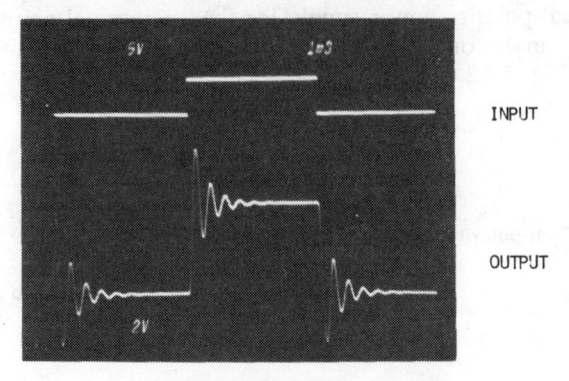

Fig. 5-13. Oscilloscope display.

Step 5

Now change the time base of the oscilloscope to 0.1 ms/division. Measure the time it takes for the oscillations to complete 1 cycle (T). From this measurement, compute the damped natural frequency (f_d) so that

$$f_d = \frac{1}{T}$$

$$= \underline{\qquad} \text{ Hz}$$

The results should be about 0.45 ms, giving a damped natural frequency of 2222 Hz.

Step 6

Now set your oscilloscope to the following settings:

- Channel 1: 0.1 V/division
- Channel 2: 0.5 V/division
- Time base: 0.1 ms/division
- Triggering: Channel 1

In addition, set your function generator for a *sine-wave* output and adjust its voltage at 0.7 volt peak-to-peak (i.e., 7 vertical divisions) and the frequency at approximately 1 kHz.

Step 7

While observing the waveform of channel 2 (the output of the active filter), slowly increase the input frequency until the peak-to-peak voltage reaches its *maximum value*. This is the natural frequency (f_n) of the system. Now, with a frequency counter, measure the frequency of the function generator and record your result:

$$f_n = \underline{\qquad} \text{ Hz}$$

The result should be approximately 2247 Hz.

Step 8

Change the sensitivity of channel 2 to 0.1 V/division. Increase the input frequency further until the peak-to-peak voltage displayed on channel 2 drops down to 0.5 V (5 vertical divisions). At this point, the output voltage will be approximately 0.707 times the input voltage, or 3 dB less than the input voltage. The frequency at which this occurs is the *3-dB bandwidth* of the system. Using a frequency counter, measure the input frequency and record your result:

$$f_{3\,dB} = \underline{\qquad} \text{ Hz}$$

Step 9

Now compare your experimental results for ζ (Step 4), f_d (Step 5), f_n (Step 7), and $f_{3\,dB}$ (Step 8) with the corresponding parameters calculated from the equations given in the "Design Basics" section of this experiment, summarizing your results in Table 5-1.

Table 5-1.

Measured Parameter	Expected Value	Experimental Value
ζ	0.146	
f_n	2322 Hz	
f_d	2297 Hz	
$f_{3\,dB}$	3554 Hz	

If all of your values are more than 5% to 10% away from the expected values, the chances are that one of the resistors and/or capacitors is significantly different when compared with its nominal value. If only one of the values is out of line, you probably made a measurement error.

CHAPTER 6

Digital Frequency Synthesizers

INTRODUCTION

Digital frequency synthesizers are phase-locked-loop systems that produce a wide range of output frequencies depending on the setting of a programmable counter. Present-day amateur, Citizens band, and aircraft communications systems frequently use some type of frequency synthesis. This chapter is divided into two basic parts. First, the basic operation of the phase-locked-loop synthesizer is discussed, including techniques used in communications systems. Second, circuits using both TTL and CMOS integrated circuits as divide-by-N counters are presented.

OBJECTIVES

At the completion of this chapter, you will be able to do the following:

- Explain the general principle of digital frequency synthesizers.
- By means of block diagrams, explain the difference between heterodyne-down conversion and prescaling-type synthesizers.
- Describe the method for designing the synthesizer loop filter.
- Describe several crystal-controlled TTL and CMOS frequency reference circuits.
- Describe several TTL and CMOS fixed and programmable counter circuits.
- Understand the use of thumbwheel switches.

THE BASIC SYNTHESIZER

Basically, a *frequency synthesizer is a frequency source whose output is an integer multiple of an input reference frequency.* As shown in Fig. 6-1, the basic frequency synthesizer is formed from a phase-locked loop by breaking the connection between the vco and the phase detector with a divide-by-N counter. Compared with the basic phase-locked-loop diagram of Fig. 1-1, the synthesizer phase detector produces an average voltage that is proportional to the phase difference between the input reference frequency, f_{REF}, and the output frequency of the divide-by-N counter, f_o/N. The counter, usually controlled by thumbwheel switches, generates a single output pulse for every N input pulses. The phase detector output voltage, after filtering, controls the output frequency of the vco (f_o) which is equal to N times the input reference during phase-lock. In addition, the input frequency will be equal to the output frequency of the divide-by-N counter, except for a finite phase difference. As before, the phase detector, loop filter, and vco make up the *forward* path of the loop, while the divide-by-N counter now constitutes the *feedback* path.

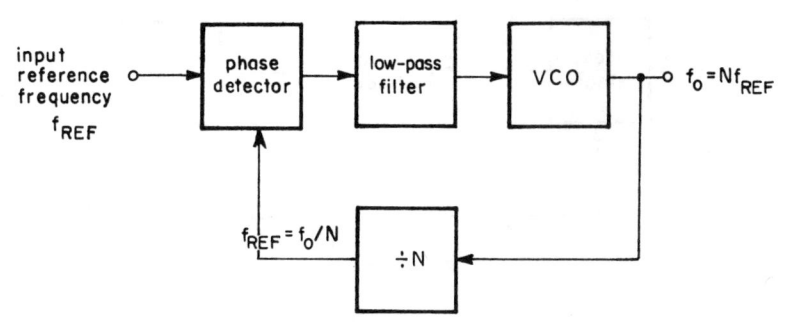

Fig. 6-1. Frequency synthesizer block diagram.

PRACTICAL SYNTHESIZERS

In communications systems, practical synthesizers generally have the output frequency operating in the 30- to 300-MHz range (vhf and uhf). Since these frequencies are beyond the maximum operating ranges of most TTL and CMOS divide-by-N counters, several techniques are used to reduce some of the frequencies generated within the loop.

First, the input reference (f_{REF}) to the phase comparator is usually less than 10 kHz. The actual frequency reference, which is a stable, crystal-controlled oscillator, usually operates from 1 to 10 MHz (f_x). Therefore, a divide-by-N counter is placed between the crystal os-

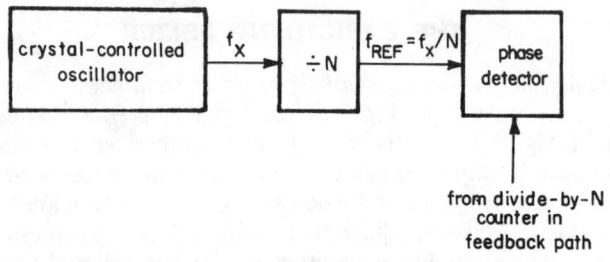

Fig. 6-2. Divide-by-N counter block diagram.

cillator and the phase detector input, as shown in Fig. 6-2. For example, if a 1-MHz, crystal-controlled oscillator is used, and the input reference frequency is to be 10 kHz, then we will need a counter that divides the 1-MHz signal by 100 to give the required 10-kHz input reference.

A second technique, known as *heterodyne-down conversion,* is illustrated in the block diagram of Fig. 6-3. A second crystal-controlled oscillator, called an *offset,* or *local* oscillator, is fed to a mixer stage where it is mixed with the output frequency of the vco (also

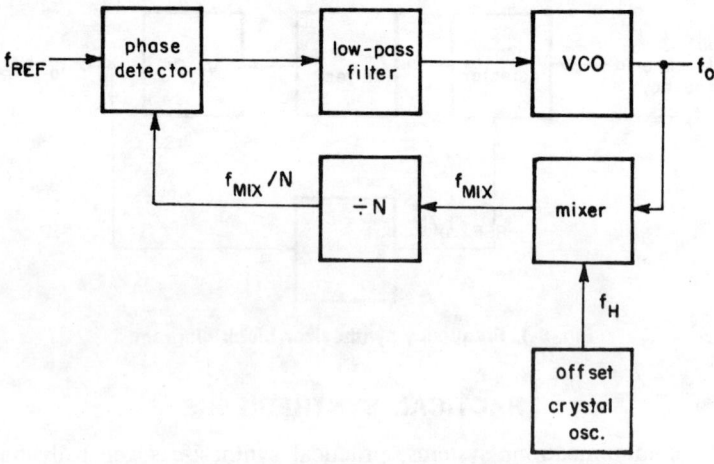

Fig. 6-3. Heterodyne-down conversion block diagram.

the output frequency of the synthesizer, f_o). The resultant output is the difference of the two inputs ($f_o - f_H$), which we will call f_{MIX}. This difference is then fed to the divide-by-N counter, whose maximum and minimum values are given by the relations:

$$N_{max} = \frac{f_{MIX(max)}}{f_{REF}} = \frac{f_{o(max)} - f_H}{f_{REF}} \quad \text{(Eq. 6-1)}$$

and,

$$N_{min} = \frac{f_{MIX(min)}}{f_{REF}} = \frac{f_{o(min)} - f_H}{f_{REF}} \qquad \text{(Eq. 6-2)}$$

To illustrate this technique, consider the following example:

Example 1

Two hundred fm broadcast channels are to be equally spaced 100 kHz apart in the range of 88 to 108 MHz. Because the fm receiver uses a 10.7-MHz intermediate frequency, the output frequency of the synthesizer must tune from 98.7 to 118.7 MHz. Using a 1-MHz crystal-controlled reference and a 98-MHz local oscillator, determine the proper values of the two divide-by-N counters, N_1 and N_2, shown in Fig. 6-4.

For a channel spacing of 100 kHz (f_{REF}), the 1-MHz reference oscillator frequency must be reduced by a factor of 10, so that $N_1 = 10$. Using Equation 6-1,

$$N_{2(max)} = \frac{f_{o(max)} - f_H}{f_{REF}}$$
$$= \frac{118.7 - 98.0}{0.1}$$
$$= 207$$

Then, from Equation 6-2,

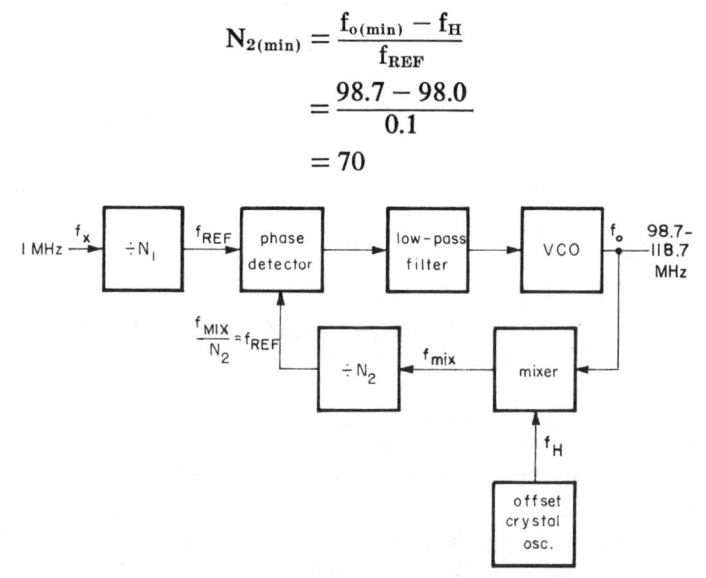

$$N_{2(min)} = \frac{f_{o(min)} - f_H}{f_{REF}}$$
$$= \frac{98.7 - 98.0}{0.1}$$
$$= 70$$

Fig. 6-4. Block diagram for heterodyne-down conversion example technique.

Consequently, we must have a divide-by-N counter capable of dividing by any integer number between 70 and 207.

Another example shows a variation of this same technique.

Example 2

Four hundred fm channels are to be equally spaced 10 kHz apart so that the receiver tunes from 144.00 to 148.00 MHz. The output of the synthesizer is multiplied by 9 and added to the 10.7-MHz intermediate frequency of the receiver. The input reference frequency is derived from a 4.551111-MHz crystal-controlled oscillator (f_x) and divided by 4096 (i.e., 2^{12}) to give 1.11111 kHz. The synthesizer must then have programmed inputs ranging from 400 (N_{min}) at 144.00 MHz to 800 (N_{max}) at 148.00 MHz. Determine the frequency of the local oscillator (f_H).

Since the receiver tunes from 144.00 to 148.00 MHz, the output frequency of the synthesizer (f_o) must be between

$$f_{o(min)} = \frac{144.00 - 10.7 \text{ MHz}}{9}$$
$$= 14.811111 \text{ MHz}$$

and,

$$f_{o(max)} = \frac{148.00 - 10.7 \text{ MHz}}{9}$$
$$= 15.255555 \text{ MHz}$$

Then, from Equation 6-1, after rearranging,

$$f_H = f_{o(max)} - N_{max}(f_{REF})$$
$$= 15.255555 \text{ MHz} - (800)(1.11111 \text{ kHz})$$
$$= 14.366667 \text{ MHz}$$

On the other hand, we could have used Equation 6-2, so that

$$f_H = f_{o(min)} - N_{min}(f_{REF})$$
$$= 14.81111 \text{ MHz} - (400)(1.11111 \text{ kHz})$$
$$= 14.366667 \text{ MHz}$$

An alternate approach to the heterodyne-down conversion technique is *prescaling*. In the synthesizer system of Fig. 6-5, a fixed counter prescales the vco down by a *constant factor* (K) to the greatest value that can be handled by the integrated circuits used for the programmable counter. Specific integrated circuits will be discussed later in this chapter. For this system, the channel spacing (f_{CH}) is given by

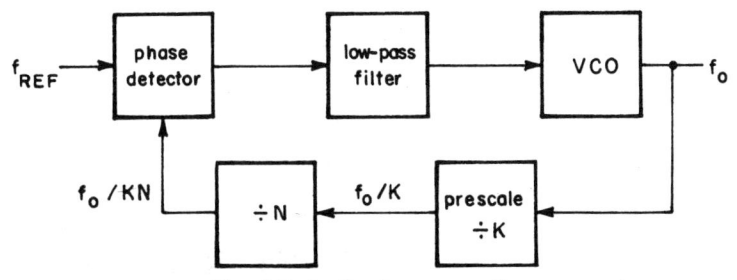

Fig. 6-5. Synthesizer prescaling.

$$f_{CH} = Kf_{REF} \quad \text{(Eq. 6-3)}$$

so that the maximum and minimum values for the programmable counter are,

$$N_{max} = \frac{f_{o(max)}}{f_{CH}} \quad \text{(Eq. 6-4)}$$

and,

$$N_{min} = \frac{f_{o(min)}}{f_{CH}} \quad \text{(Eq. 6-5)}$$

Example 3

As before, 200 fm channels are to be equally spaced 100 kHz apart in the 88- to 108-MHz range. The synthesized output frequency is to be 98.7 to 118.7 MHz, but with a reference frequency of 10 kHz derived from a 1-MHz crystal-controlled oscillator. From Equation 6-3, the divisor (K) is found,

$$\begin{aligned} K &= \frac{f_{CH}}{f_{REF}} \\ &= \frac{100 \text{ kHz}}{10 \text{ kHz}} \\ &= 10 \end{aligned}$$

Then, from Equations 6-4 and 6-5,

$$\begin{aligned} N_{max} &= \frac{f_{o(max)}}{f_{CH}} \\ &= \frac{118.7}{0.1} \\ &= 1187 \end{aligned}$$

and,

$$N_{min} = \frac{f_{o(min)}}{f_{CH}}$$
$$= \frac{98.7}{0.1}$$
$$= 987$$

Finally, the 10-kHz reference (f_{REF}) is obtained from a 1-MHz crystal-controlled oscillator using a divide-by-100 counter. The completed block diagram of the prescaled synthesizer is shown in Fig. 6-6. In comparison to the prescaling method, the reference frequency (f_{REF}) in the heterodyne-down system is equal to the channel spacing. However, the latter method requires the use of an additional crystal oscillator, which may cause unwanted spurious frequencies at the mixer output if not completely filtered.

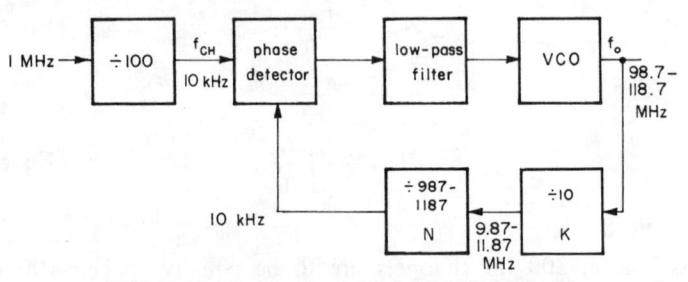

Fig. 6-6. Block diagram of prescaled synthesizer.

THE SYNTHESIZER LOOP FILTER

There is basically no difference in choosing the parameters of the loop filter between the basic phase-locked-loop system and the digital synthesizer. The only precaution being that we must now make sure that the filter gives the desired response over the entire range of the synthesizer.

The step-by-step determination of the loop filter parameters can be summarized as follows:

1. Choose the desired channel spacing (the reference oscillator input frequency, f_{REF}.
2. Calculate the range of digital division from

$$N_{max} = \frac{f_{o(max)}}{f_{REF}} \qquad \text{(Eq. 6-6)}$$

and,

$$N_{min} = \frac{f_{o(min)}}{f_{REF}} \qquad \text{(Eq. 6-7)}$$

3. Determine the vco range from

$$(2f_{o(max)} - f_{o(min)}) \leq f_{vco} \leq (2f_{o(min)} - f_{o(max)}) \quad \text{(Eq. 6-8)}$$

4. Depending on the type of filter used, choose a minimum value for the damping factor (ζ_{min}) from the transient response curves in Chapter 5.
5. Calculate the loop natural frequency (ω_n) based on the desired settling time (t_s) and the $\omega_n t_s$ product obtained from the transient response curve.
6. Calculate the minimum value of capacitance of the loop filter:

$$C_{min} = \frac{K_\phi K_p}{N_{max} R \omega_n^2} \quad \text{(for Fig. 5-2)} \quad \text{(Eq. 6-9a)}$$

$$C_{min} = \frac{K_\phi K_o}{N_{max}(R_I + R_2)\omega_n^2} \quad \text{(for Fig. 5-3)} \quad \text{(Eq. 6-9b)}$$

or,

$$C_{min} = \frac{K_\phi K_o}{N_{max} R_1 \omega_n^2} \quad \text{(for Fig. 5-4)} \quad \text{(Eq. 6-9c)}$$

7. Determine the maximum damping factor from

$$\zeta_{max} = \zeta_{min} \left(\frac{N_{max}}{N_{min}}\right)^{1/2} \quad \text{(Eq. 6-10)}$$

8. Check the transient response of ζ_{max} for compliance with initial design specifications.

Example

To illustrate this procedure, consider the following specifications:

- Synthesizer output range: 88–108 MHz
- Input reference frequency: 0.1 MHz
- Settling time: 10 ms at 10% overshoot
- Maximum overshoot: 20%

Then the following step-by-step determinations are made:

1. $f_{REF} = 0.1$ MHz
2. $N_{max} = 108/0.1 = 1080$, and $N_{min} = 88/0.1 = 880$
3. Vco range:

$$2f_{o(max)} - f_{o(min)} = (2)(108) - 88 = 128 \text{ MHz}$$

and,

$$2f_{o(min)} - f_{o(max)} = (2)(88) - 108 = 68 \text{ MHz}$$

4. Using the active filter of Fig. 5-4 and the transient response curve of Fig. 5-7, we find that a damping factor of 0.8 will give an overshoot of less than 20%.

5. Also from Fig. 5-7, the transient response will be less than 10% at $\omega_n t_s = 3.5$, so that for a settling time of 10 ms,

$$\omega_n = \frac{\omega_n t_s}{t_s} = \frac{3.5}{10 \text{ ms}}$$
$$= 350 \text{ rad/s } (55.7 \text{ Hz})$$

6. Assuming that we plan to use an MC4024 integrated circuit for the vco ($K_o = 11 \times 10^6$ rad/s/V), an MC4044 phase detector ($K_\phi = 0.12$ V/rad), and picking $R_1 = 4.7$ kΩ, then from Equation 6-9c,

$$C_{min} = \frac{K_\phi K_o}{N_{max} R_1 \omega_n^2}$$
$$= \frac{(11 \times 10^6)(0.12)}{(1080)(4.7 \text{ k}\Omega)(350)^2}$$
$$= 2.1 \text{ }\mu\text{F (use 2 }\mu\text{F)}$$

Then, by rearranging Equation 5-11, R_2 can be calculated,

$$R_2 = \frac{2\zeta_{min}}{\omega_n C_{min}}$$
$$= \frac{(2)(0.8)}{(350)(2 \text{ }\mu\text{F})}$$
$$= 2285 \text{ }\Omega \text{ (use 2.2 k}\Omega\text{)}$$

7.
$$\zeta_{max} = \zeta_{min} \left(\frac{N_{max}}{N_{min}}\right)^{1/2}$$
$$= (0.8)\left(\frac{1080}{880}\right)^{1/2}$$
$$= 0.89$$

8. From Fig. 5-7, we see that for a damping factor of 0.89 the transient response will have an overshoot of less than 10% within 10 ms.

FREQUENCY REFERENCE CIRCUITS

In order for a phase-locked-loop synthesizer system to function accurately, it must possess a *stable* reference frequency. That is, the reference frequency must be held essentially constant over wide variations in ambient temperature, circuit loads, and power-supply voltage. To meet these objectives, some form of crystal-controlled oscillator is used. In this section, several popular TTL and CMOS oscillators are discussed.

TTL Oscillators

One of the most reliable TTL oscillator circuits is the one shown in Fig. 6-7, used for crystal frequencies between 1 and 10 MHz. The two 470-Ω resistors help to assure that the 7400 NAND gates (U1 and U2) operate in a somewhat linear fashion. This guarantees reliable starting when the power is applied, in addition to producing a temperature stabilizing effect. The variable capacitor (C) is used to accurately adjust the oscillator frequency against a known standard such as WWV. Reducing the value of C will increase the output frequency, and vice versa. The remaining NAND gate (U3) serves as a buffer. Other variations of this circuit are shown in Fig. 6-8. In all of these circuits, the crystal (Y1) operates in a *series-resonant mode*. Consequently, the standard AT-cut crystal is the best choice.

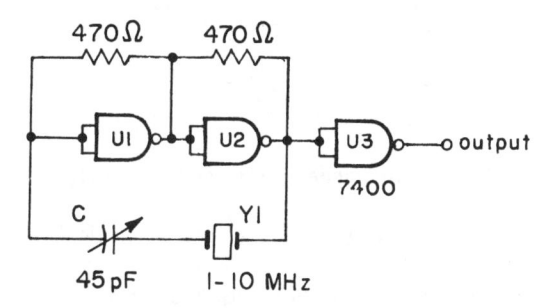

Fig. 6-7. Crystal-controlled TTL oscillator.

For higher crystal frequencies, the MC4024 dual voltage-controlled multivibrator (refer to Figs. 4-2 and 4-3) can be wired to function as a crystal oscillator. As shown in Fig. 6-9, a crystal is used in place of the external frequency-determining capacitor. The output frequency is then adjusted against a known standard by adjusting the 5-kΩ potentiometer which, as part of a voltage divider, controls the dc input voltage. The maximum operating frequency is 25 MHz.

CMOS Oscillators

Using CMOS devices,[*] a popular circuit using NOR gates is shown in Fig. 6-10, which is useful up to 4 MHz. Although the crystal is in a *parallel-resonant mode,* the AT-cut–type crystal is still preferred.

---

[*]For further information on CMOS crystal oscillators, see RCA Application Note ICAN-6086, "Timekeeping Advances Through COS/MOS Technology," by S. S. Eaton.

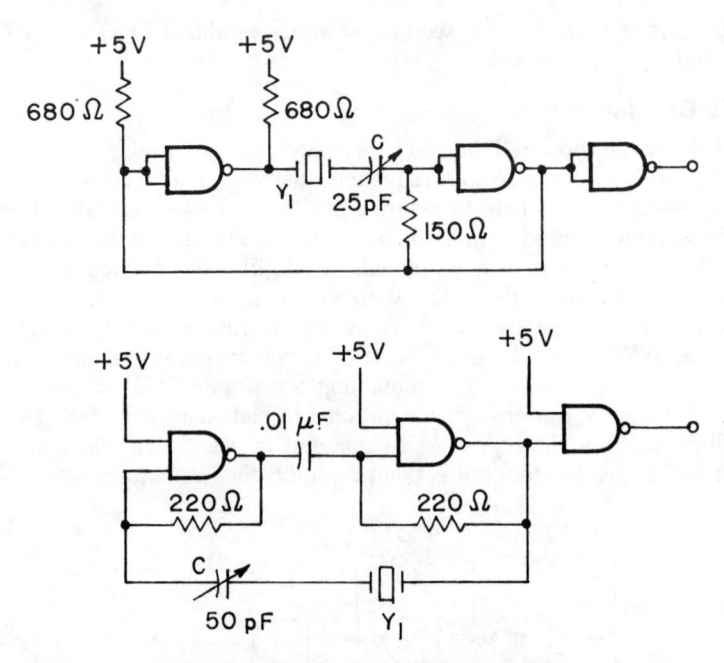

Fig. 6-8. Additional TTL crystal oscillators.

Instead of type 4001 NOR gates, an alternative device is the 4060, which is a combination oscillator and binary divider (Fig. 6-11). In addition to the fundamental output frequency (f_o), determined by the crystal, the device also divides this frequency in binary multiples from 16 to 16,384. As shown in Fig. 6-12, the oscillator circuit is capable

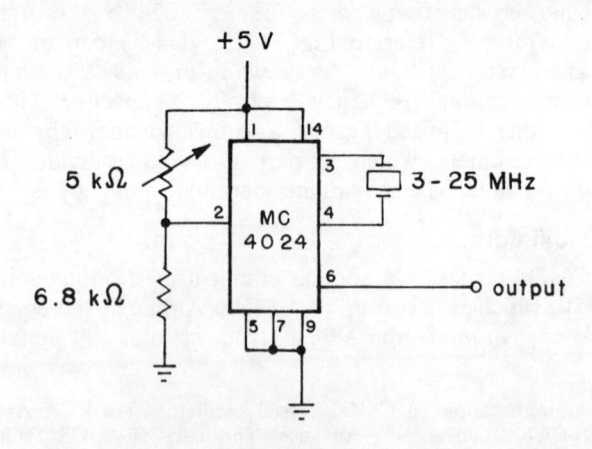

Fig. 6-9. MC4024 crystal oscillator.

88

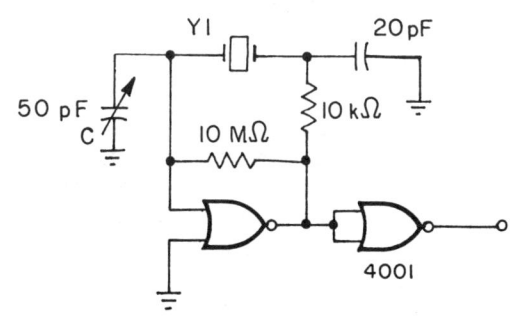

Fig. 6-10. CMOS crystal oscillator.

Fig. 6-11. Pin diagram of the MC4060 CMOS oscillator/binary divider.

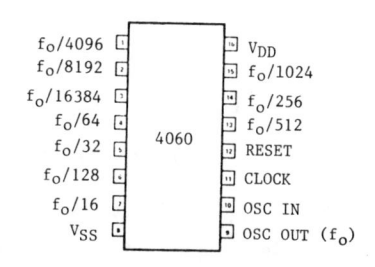

$f_o/4096$ — V_{DD}
$f_o/8192$ — $f_o/1024$
$f_o/16384$ — $f_o/256$
$f_o/64$ — $f_o/512$
$f_o/32$ — 4060 — RESET
$f_o/128$ — CLOCK
$f_o/16$ — OSC IN
V_{SS} — OSC OUT (f_o)

of generating a total of 10 output frequencies. However, there is no output available for divisions by 2, 4, 8, or 2048. At a supply voltage ($+V_{DD}$) of 5 volts, the maximum operating frequency possible is 1.75 MHz; at $V_{DD} = 10$ volts, it is 4 MHz.

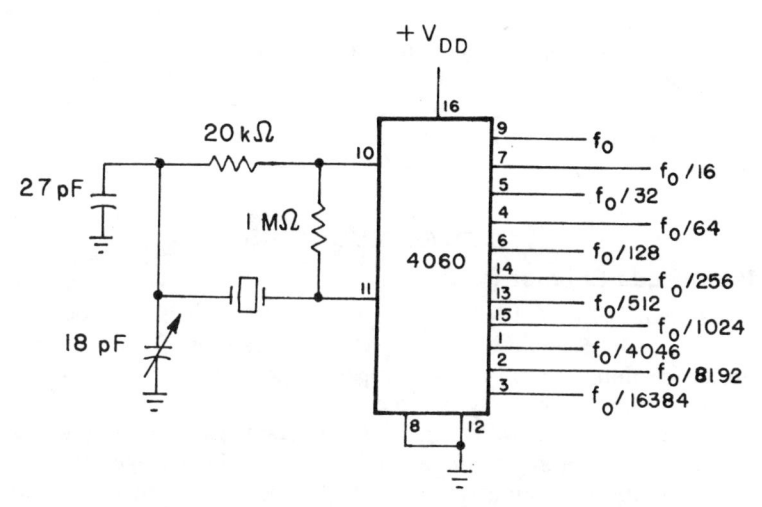

Fig. 6-12. Circuit diagram of the MC4060 oscillator.

DIVIDE-BY-N COUNTERS

Briefly stated, a *divide-by-N counter is a digital logic circuit that produces a single output pulse for every N input pulses, where N is an integer.* It is referred to as the *modulus,* or *modulo,* of the counter. In an earlier section, we learned that the reference frequency of the phase-locked-loop synthesizer is generally some fraction of the main oscillator frequency. For example, an input reference frequency of 833.333 Hz derived from a 1-MHz master oscillator requires a divide-by-N counter having a modulus of 1200. In addition, the feedback element of a synthesizer is also a divide-by-N counter, so that the output frequency of the synthesizer is N *times* the input reference.

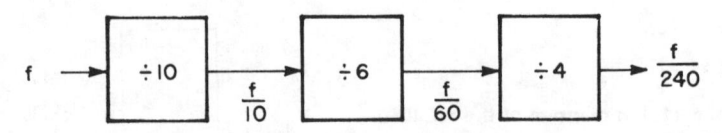

$$f \longrightarrow \boxed{\div 10} \xrightarrow{\frac{f}{10}} \boxed{\div 6} \xrightarrow{\frac{f}{60}} \boxed{\div 4} \longrightarrow \frac{f}{240}$$

Fig. 6-13. Cascading of fixed divide-by-N counters.

In general, the divide-by-N counter associated with the input reference has a *fixed* modulus, so that the resultant input reference is always the same. On the other hand, the divide-by-N counter in the synthesizer feedback loop is usually *programmable,* so that the modulus can be manually changed to any number. Since TTL and CMOS divide-by-N counters, both fixed and programmable, are somewhat different in their operation, they will be treated separately in this chapter.

Fixed-modulus–type counters are normally *cascaded,* or connected in sequence so that the output of the first counter is the input to the second, etc., as illustrated in Fig. 6-13. In this manner, the resultant count will be increased. The *total modulus* will be the product of the individual modulos.

TTL FIXED COUNTERS

7490 Decade Counter ($\div 10$)

As shown in Fig. 6-14, the 7490 integrated circuit is a divide-by-2 and a divide-by-5 counter in a single 14-pin package. However both counter sections may be used together to form a divide-by-10 counter, which is frequently the case.

Fig. 6-15 shows the 7490 wired as a divide-by-5 counter, while Fig. 6-16 shows the required connections for divide-by-10 operation. In either case, the maximum typical input frequency is limited to about 30 MHz.

Fig. 6-14. Pin diagram of the 7490 decade counter.

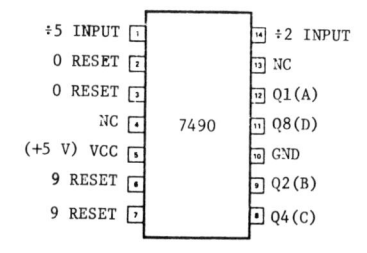

```
÷5 INPUT  [1]        [14] ÷2 INPUT
0 RESET   [2]        [13] NC
0 RESET   [3]        [12] Q1(A)
NC        [4] 7490   [11] Q8(D)
(+5 V) VCC[5]        [10] GND
9 RESET   [6]        [9]  Q2(B)
9 RESET   [7]        [8]  Q4(C)
```

7492 Divide-by-12 Counter

As shown in Fig. 6-17, the 7492 is a divide-by-2 and a divide-by-6 counter in a single 14-pin package. Although both sections may be used together as a divide-by-12 counter, the 7492 has its major use as a divide-by-6 counter using the connections of Fig. 6-18, which is usable up to about 18 MHz.

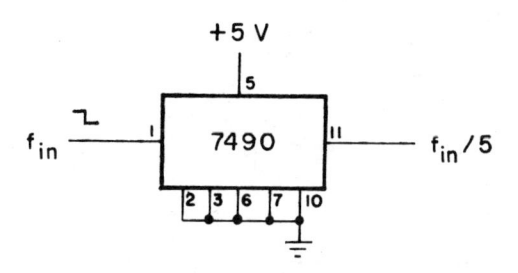

Fig. 6-15. The 7490 connected as a divide-by-5 counter.

Example

Using a crystal-controlled reference oscillator, describe a circuit using 7490 and 7492 counters that will divide a 6-MHz input frequency down to 20 kHz.

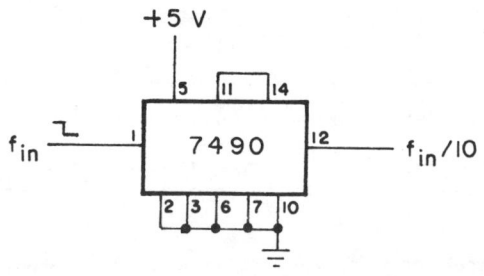

Fig. 6-16. The 7490 connected as a divide-by-10 counter.

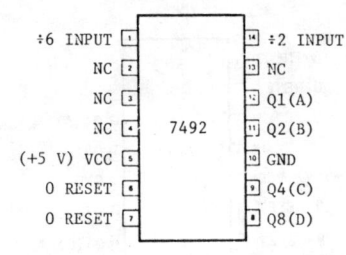

```
÷6 INPUT  1      14  ÷2 INPUT
      NC  2      13  NC
      NC  3      12  Q1(A)
      NC  4  7492 11  Q2(B)
(+5 V) VCC  5      10  GND
   0 RESET  6       9  Q4(C)
   0 RESET  7       8  Q8(D)
```

Fig. 6-17. Pin diagram of the 7492 divide-by-12 counter.

To go from 6 MHz down to 20 kHz, we must have a series of counters which, when cascaded, will divide the input by 300. Using only the 7490 and 7492 counters, we would require sections that will divide by 6, 5, and then 10 (or any other order), as shown in Fig. 6-19.

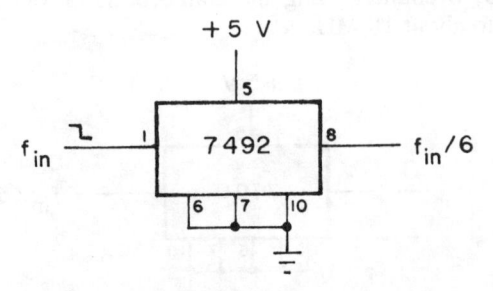

Fig. 6-18. The 7492 connected as a divide-by-6 counter.

CMOS FIXED COUNTERS
4017 and MM4617 Decade Counters (÷10)

The 4017 (Fig. 6-20) (MM4617 by National Semiconductor) is a 5-stage Johnson decade counter which can be used to divide an input frequency by 10 using the circuit of Fig. 6-21.

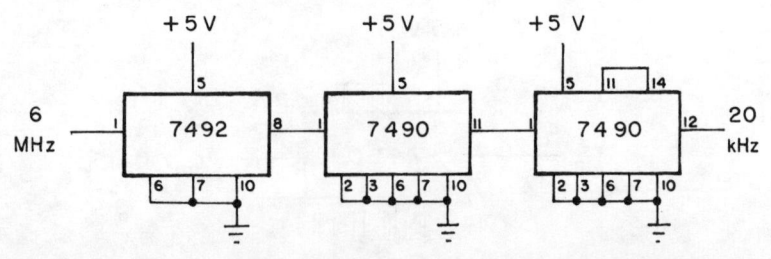

Fig. 6-19. Circuit for dividing 6 MHz down to 20 kHz using cascaded 7490 and 7492 counters.

Fig. 6-20. Pin diagram of the 4017 decade counter.

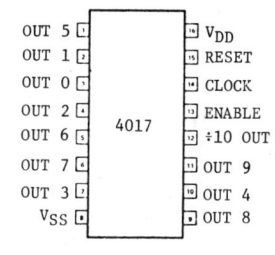

```
OUT 5  -|  1    16 |- V_DD
OUT 1  -|  2    15 |- RESET
OUT 0  -|  3    14 |- CLOCK
OUT 2  -|  4  4017 13 |- ENABLE
OUT 6  -|  5    12 |- ÷10 OUT
OUT 7  -|  6    11 |- OUT 9
OUT 3  -|  7    10 |- OUT 4
V_SS   -|  8     9 |- OUT 8
```

4018 Presettable Divide-by-N Counter

The 4018 (Fig. 6-22) is a specialized counter chip, as it can be connected to divide an input frequency by any integer from 2 to 10. Consequently, the 4018 is one of the most versatile devices for fixed modulos. For use in synthesizers, the divide-by-2, -4, -6, and -10 configurations are the most useful, and are shown in Fig. 6-23. The maxi-

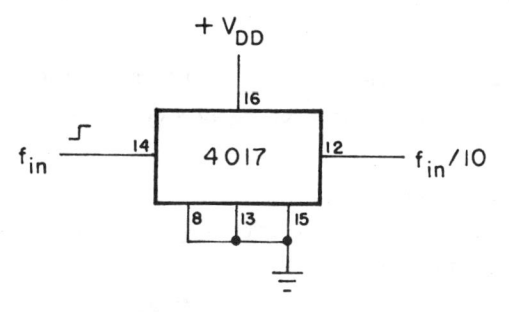

Fig. 6-21. The 4017 connected as a divide-by-10 counter.

mum typical input frequency for both the 4017 and 4018 is 2.5 MHz with a supply voltage ($+V_{DD}$) of 5 volts; for $V_{DD} = 10$ volts, it is 5 MHz.

Fig. 6-22. Pin diagram of the 4018 presettable divide-by-N counter.

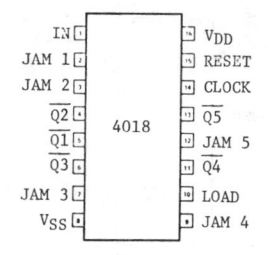

```
IN    -|  1    16 |- V_DD
JAM 1 -|  2    15 |- RESET
JAM 2 -|  3    14 |- CLOCK
Q2    -|  4  4018 13 |- Q5
Q1    -|  5    12 |- JAM 5
Q3    -|  6    11 |- Q4
JAM 3 -|  7    10 |- LOAD
V_SS  -|  8     9 |- JAM 4
```

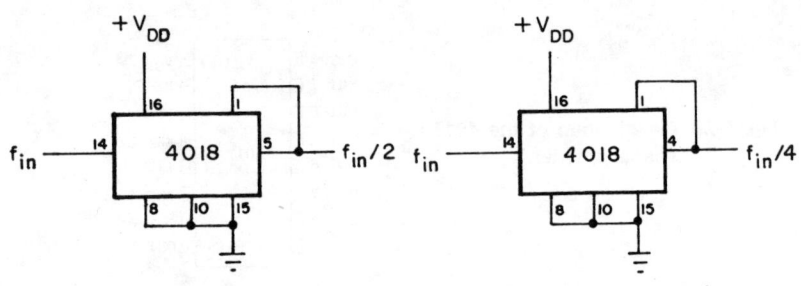

(A) Divide-by-2 connections.

(B) Divide-by-4 connections.

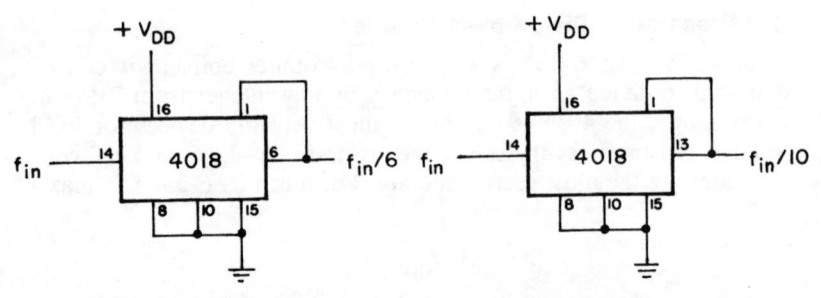

(C) Divide-by-6 connections.

(D) Divide-by-10 connections.

Fig. 6-23. Connections for the 4018 divide-by-N counter.

74C90 Decade Counter ($\div 10$)

The 74C90 is a pin-for-pin CMOS equivalent of the 7490 TTL decade counter (see Fig. 6-14).

TTL PROGRAMMABLE COUNTERS

74192 Counter

Perhaps the most frequently used TTL programmable counter is the 74192 chip. As shown in the pin diagram of Fig. 6-24, this device has four data inputs: A, B, C, and D (corresponding to a *binary*

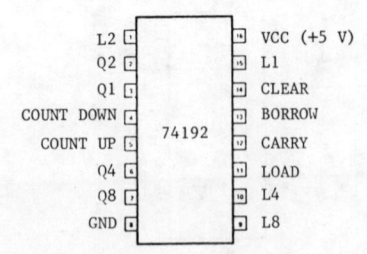

Fig. 6-24. Pin diagram of the 74192 programmable counter.

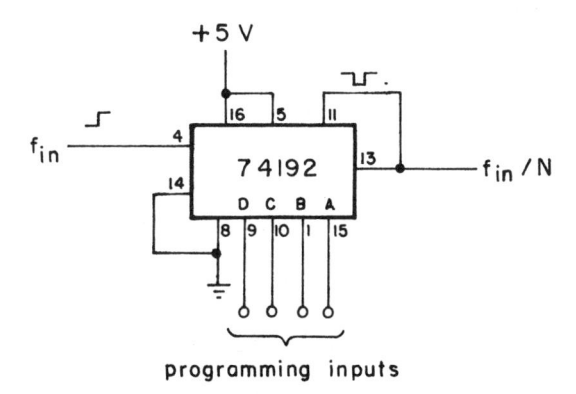

Fig. 6-25. Programming the 74192 counter.

weighting of 1, 2, 4, and 8, respectively) for programming the desired modulus. Fig. 6-25 shows how the 74192 is used to divide an input frequency by any number from 1 to 10. The modulus (N) is determined by the 4-bit input data, DCBA, according to Table 6-1.

Table 6-1.

N	D	C	B	A
1	0	0	0	1
2	0	0	1	0
3	0	0	1	1
4	0	1	0	0
5	0	1	0	1
6	0	1	1	0
7	0	1	1	1
8	1	0	0	0
9	1	0	0	1
10	1	0	1	0

When the LOAD input (pin 11) is briefly brought low (logic 0), the counter is preset with the 4-bit binary number present at the data inputs. The counter then counts *downward* on each positive edge of the input waveform. For example, if the binary number DCBA = 0110 (i.e., the decimal number 6) were present at the data inputs, the 74192 counts backward: 6, 5, 4, . . ., until it reaches 0, at which point the counter is again parallel-loaded with the 4-bit code 0110. The cycle is then repeated. Actually, the 74192 is capable of dividing by any number up to 15; however, this makes no sense when we cascade several counters to form multidecade divide-by-N counters. For the 74192, the maximum input frequency is typically 32 MHz.

Compared with the fixed-modulus counters, the 74192 is *unit-decade-cascadable,* a feature which makes it extremely useful for frequency synthesizers. Suppose we want to divide the input frequency by 596. We would then cascade three 74192 counters so that the first one (*units* decade) is loaded with a "6," the second one (*tens* decade) with a "9," and the third one (*hundreds* decade) with a "5" so that the three decades are conveniently separated into individual groups.

With the circuit of Fig. 6-26, three 74192 counters are unit-decade-cascaded and programmed to divide the input frequency by 596. The units decade begins by counting down from 6, so that the sequence goes: 596, 595, 594, . . ., etc. When the units count reaches 0, it then *borrows* from the tens decade counter so that the next count will be 589 and the process continues. When the down count reaches 500, the tens counter borrows from the hundreds counter so that the next count is 499. When the hundreds decade reaches 0 (a down count of 099), *it does not reload,* since the other two counters have not yet reached 0. Only when all three counters are at a down count of 0 are the counters then again parallel-loaded with the three, 4-bit binary numbers.

MC4016 Counter

Another frequently used TTL programmable counter is the MC-4016 (Motorola) (Fig. 6-27), and is similar in its operation to the 74192. *This device should not be confused with the 4016 CMOS quad bilateral switch.*

Fig. 6-28 illustrates the method for cascading three MC4016 counters to divide an input frequency by any integer from 1 to 999.

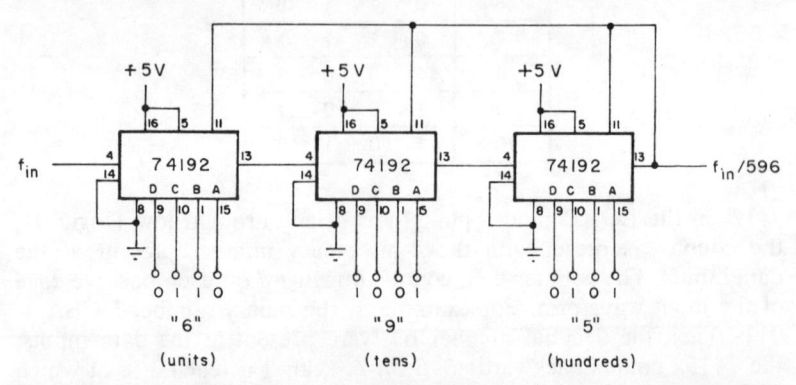

Fig. 6-26. Three 74192 counters unit-decade-cascaded and programmed to divide by 596.

Fig. 6-27. Pin diagram of the MC4016 programmable counter.

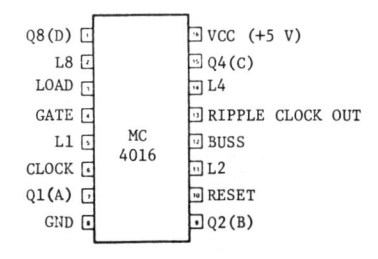

```
Q8(D)  1      16  VCC (+5 V)
L8     2      15  Q4(C)
LOAD   3      14  L4
GATE   4  MC  13  RIPPLE CLOCK OUT
L1     5  4016 12  BUSS
CLOCK  6      11  L2
Q1(A)  7      10  RESET
GND    8       9  Q2(B)
```

Like the 74192, the MC4016 counts down to 0 from its preset 4-bit binary number; however, the maximum input frequency is typically 8 MHz, or one-fourth that of the 74192.

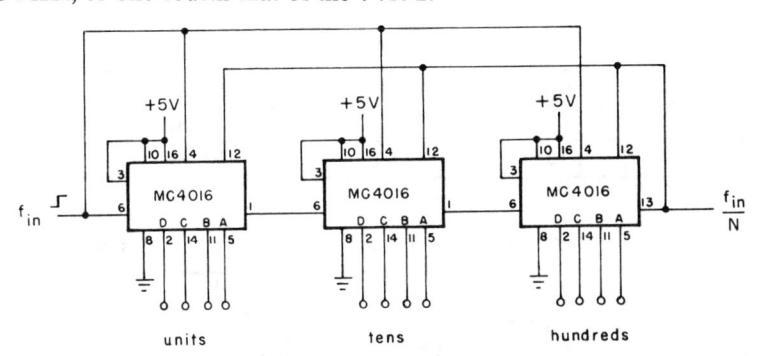

Fig. 6-28. Three MC4016 counters cascaded to divide by any integer from 1 to 999.

CMOS PROGRAMMABLE COUNTERS

4018 Counter

Although the 4018 counter (see Fig. 6-22) can be wired for multi-decade division, the scheme nevertheless requires additional inverters, NOR gates, and D-type flip-flops, all of which increase the total component count. An RCA applications note, ICAN-6498 entitled, *Design of Fixed and Programmable Counters Using the RCA CD4018 COS/MOS Presettable Divide-by-"N" Counter* explains the method involved. The following CMOS devices described are better suited for programmable division.

40192 and 74C192 Counters

The 40192 (RCA) and the 74C192 (National Semiconductor) counters are both direct pin-for-pin CMOS equivalents of the 74192 TTL device (see Fig. 6-24). The typical maximum input frequencies are summarized in Table 6-2.

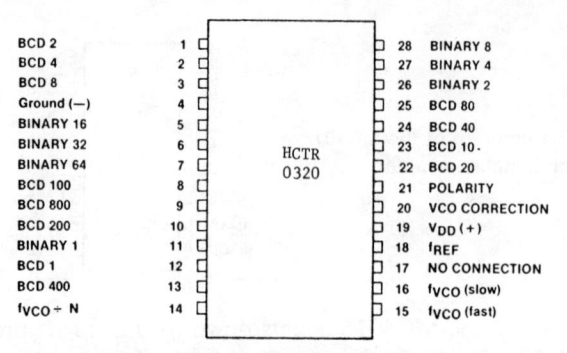

```
BCD 2         1          28  BINARY 8
BCD 4         2          27  BINARY 4
BCD 8         3          26  BINARY 2
Ground (—)    4          25  BCD 80
BINARY 16     5          24  BCD 40
BINARY 32     6   HCTR   23  BCD 10
BINARY 64     7   0320   22  BCD 20
BCD 100       8          21  POLARITY
BCD 800       9          20  VCO CORRECTION
BCD 200      10          19  $V_{DD}$ (+)
BINARY 1     11          18  $f_{REF}$
BCD 1        12          17  NO CONNECTION
BCD 400      13          16  $f_{VCO}$ (slow)
$f_{VCO} \div N$  14     15  $f_{VCO}$ (fast)
```

Fig. 6-29. Pin diagram of the HCTR 0320 digital frequency synthesizer.

Table 6-2.

V_{DD}	40192	74C192
+5 V	3 MHz	4 MHz
+10 V	7 MHz	10 MHz

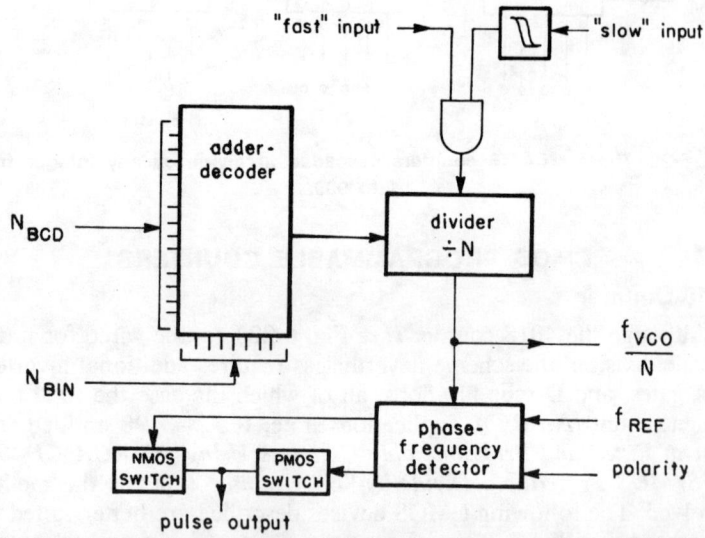

Fig. 6-30. Block diagram of the HCTR 0320 digital frequency synthesizer.

HCTR 0320 Digital Frequency Synthesizer

The HCTR 0320 (Hughes) is a CMOS device which contains both a programmable divider and a phase/frequency detector in a 28-pin package, as shown in Fig. 6-29. The general operation of this device is best explained by the block diagram of Fig. 6-30.

Adder/Decoder

This block adds a three-digit *bcd* number (N_{bcd}) ranging from 0 to 999, to a 7-bit *binary* number (N_{bin}) ranging from 0 to 127 (i.e., 0000000 to 1111111). The total modulus is the sum, $N = N_{bcd} + N_{bin}$, which can range from 3 to 1023.

Programmable Divider

This block outputs a waveform whose frequency is 1/N of the input, with a duty cycle of 1/N. In addition, there are two types of inputs. The "fast" input at pin 15 is the only TTL-compatible input and should be used for inputs having fast rise and fall times, or when maximum speed is required. Otherwise, for inputs with slow rise and fall times (e.g., a sine wave), the "slow" input (pin 16) uses an internal Schmitt trigger for proper conditioning. The characteristics of these two inputs are summarized in Table 6-3.

Table 6-3. Input Characteristics of HCTR 0320

Input	Maximum Frequency $V_{DD} = 5$ V	$V_{DD} = 10$ V	10%–90% Rise and Fall Time $V_{DD} = 5$ V	$V_{DD} = 10$ V
"Fast" (Pin 15)	5 MHz	10 MHz	100 ns	50 ns
"Slow" (Pin 16)	2.5 MHz	5 MHz	No Limit	No Limit

Phase/Frequency Detector

This block compares the programmable divider output (f_{vco}/N) with the external frequency reference f_{REF}, generating the proper error signal to the loop filter. When the error signal goes from the floating state (NMOS and PMOS switches "off") to either the positive supply (V_{DD}) or ground (V_{SS}), the output at pin 20 is a pulse whose width is proportional to the time difference between the leading edges of f_{vco} and f_{REF}. The polarity input (pin 21) should be tied to V_{DD} if the vco correction output voltage should decrease to cause an increase in the vco frequency.

4522 Counter

The 4522 counter (Fig. 6-31) is a 16-pin device that operates like the 74C192. As shown in the 1-decade circuit of Fig. 6-32, the 4522 counts down on positive leading edges of the input signal until the count reaches 0. As soon as it reaches 0, the counter is then loaded with the 4-bit number that is present at the data input. As shown in the 2-decade circuit of Fig. 6-33, any number of 4522 counters may be cascaded. In this case, the circuit will divide the input frequency

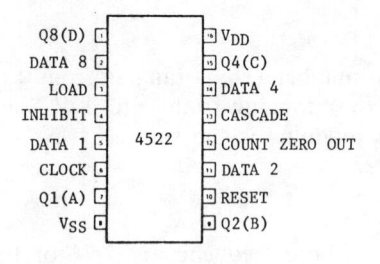

```
Q8(D)   ▢1       16▢ V_DD
DATA 8  ▢2       15▢ Q4(C)
LOAD    ▢3       14▢ DATA 4
INHIBIT ▢4  4522 13▢ CASCADE
DATA 1  ▢5       12▢ COUNT ZERO OUT
CLOCK   ▢6       11▢ DATA 2
Q1(A)   ▢7       10▢ RESET
V_SS    ▢8        9▢ Q2(B)
```

Fi. 6-31. Pin diagram of the 4522 programmable counter.

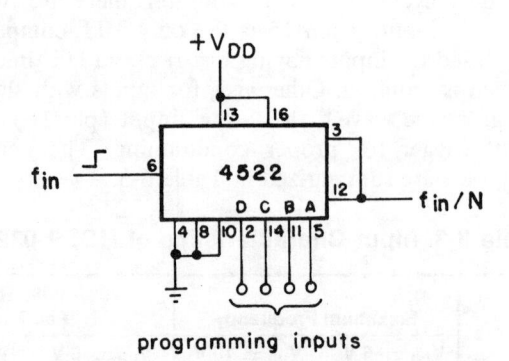

Fig. 6-32. Programming the 4522 counter.

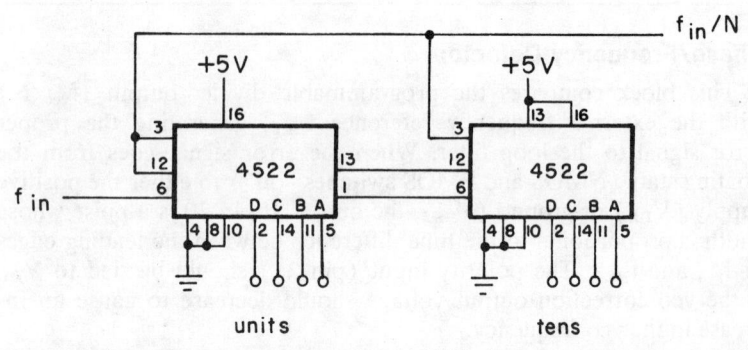

Fig. 6-33. Two 4522 counters cascaded to divide by any integer from 1 to 99.

by any integer from 1 to 99. At 5 volts, the 4522 has a maximum frequency of 1 MHz; at 10 volts, it is 2.5 MHz.

PROGRAMMING SWITCHES

All of the programmable counters described are programmed by loading either a logic 0 or a logic 1 at the desired inputs. This can be

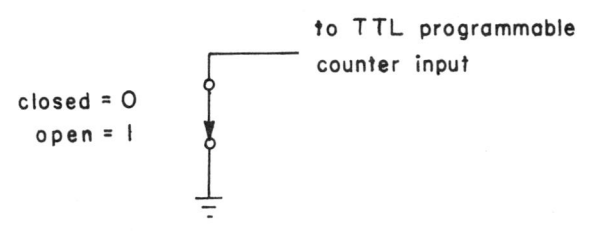

```
                    to TTL programmable
                    counter input
   closed = 0
   open = 1
```

Fig. 6-34. Switch circuit for programming TTL programmable counters.

accomplished simply with a series of slide or toggle switches. When using TTL devices, this is accomplished by either grounding (for a logic 0) or ungrounding (for a logic 1) the given input line, as illustrated by the spst switch circuit of Fig. 6-34. However, when using TTL devices, it is a good design practice to return the ungrounded TTL input to +5 volts, as shown in Fig. 6-35, in order to preserve its noise immunity ability.

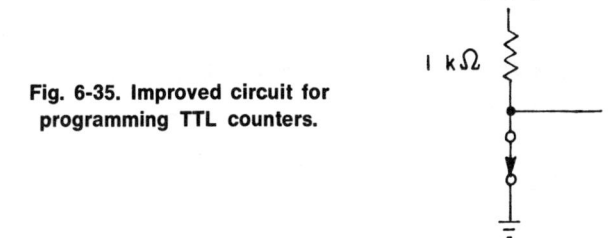

Fig. 6-35. Improved circuit for programming TTL counters.

+5 V
1 kΩ

On the other hand, *inputs of CMOS devices must be terminated either at logic 0 or logic 1; they cannot be left floating!* Otherwise, a floating input can randomly determine what the circuit will do, in addition to drastically increasing the supply current, possibly destroying the device. Consequently, all inputs of CMOS devices must follow the scheme shown in Fig. 6-36.

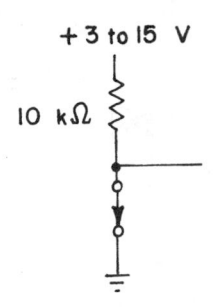

Fig. 6-36. Switching method that must be used for programming CMOS counters.

+3 to 15 V
10 kΩ

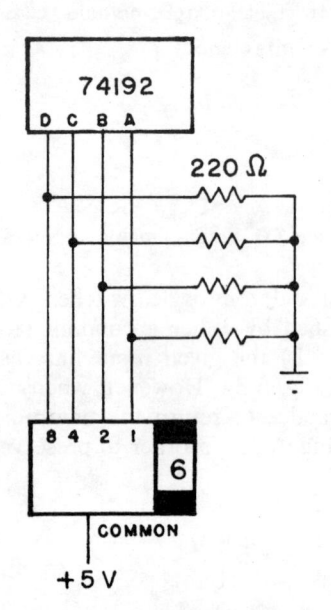

	D	C	B	A
N	(8)	(4)	(2)	(1)
0	0	0	0	0
1	0	0	0	X
2	0	0	X	0
3	0	0	X	X
4	0	X	0	0
5	0	X	0	X
6	0	X	X	0
7	0	X	X	X
8	X	0	0	0
9	X	0	0	X

X = output line connected to common

0 = open circuit

Fig. 6-37. Bcd thumbwheel switch circuit.

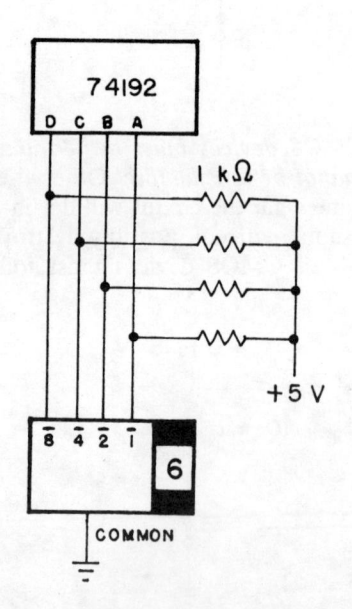

	D	C	B	A
N	(8̄)	(4̄)	(2̄)	(1̄)
0	X	X	X	X
1	X	X	X	0
2	X	X	0	X
3	X	X	0	0
4	X	0	X	X
5	X	0	X	0
6	X	0	0	X
7	X	0	0	0
8	0	X	X	X
9	0	X	X	0

X = output line connected to common

0 = open circuit

Fig. 6-38. Bcd-complement thumbwheel switch circuit.

For practical synthesizers, it is more convenient to use *thumbwheel* switches for programming each decade. A thumbwheel switch is a binary-coded, 10-position switch, and displays a single decimal digit while at the same time supplying the bcd equivalent of that digit between the "common" and the four switch leads. It is easier to use these switches to display the final output frequency directly without the use of bcd decoder/driver-LED display circuits. A 1-decade circuit using a bcd thumbwheel switch and a 74192 counter is shown in Fig. 6-37. On the other hand, it may be easier to use bcd *complement*-type thumbwheel switches to unground the proper input automatically, as shown in the 1-decade circuit of Fig. 6-38.

AN INTRODUCTION TO THE EXPERIMENTS

The following experiments are designed to demonstrate the operation of several fixed and programmable TTL and CMOS divide-by-N counters, as well as the operation of a simple digital frequency synthesizer. The experiments that you will perform can be summarized as follows:

Experiment No.	Purpose
1	Demonstrate the operation of the 7490 and 7492 TTL counters.
2	Demonstrate the capabilities of the 4017 CMOS decade counter.
3	Demonstrate the operation of the 74192 programmable TTL decade counter.
4	Demonstrate the operation of two 74192 programmable TTL counters connected in unit-cascade.
5	Demonstrate the operation of the HCTR 0320 CMOS programmable counter.
6	Demonstrate the operation of a 3-decade TTL frequency synthesizer.

EXPERIMENT NO. 1

Purpose

The purpose of this experiment is to demonstrate the operation of some TTL fixed-modulus counters, using the 7490 and 7492 integrated circuits.

103

Pin Configuration of Integrated-Circuit Chips (Fig. 6-39)

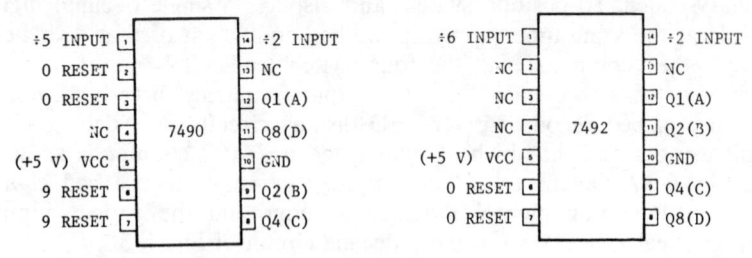

```
÷5 INPUT   [1]         [14] ÷2 INPUT        ÷6 INPUT  [1]         [14] ÷2 INPUT
0 RESET    [2]         [13] NC              NC        [2]         [13] NC
0 RESET    [3]   7490  [12] Q1(A)           NC        [3]   7492  [12] Q1(A)
NC         [4]         [11] Q8(D)           NC        [4]         [11] Q2(B)
(+5 V) VCC [5]         [10] GND             (+5 V) VCC[5]         [10] GND
9 RESET    [6]         [9]  Q2(B)           0 RESET   [6]         [9]  Q4(C)
9 RESET    [7]         [8]  Q4(C)           0 RESET   [7]         [8]  Q8(D)
```

Fig. 6-39.

Schematic Diagram of Circuit (Fig. 6-40)

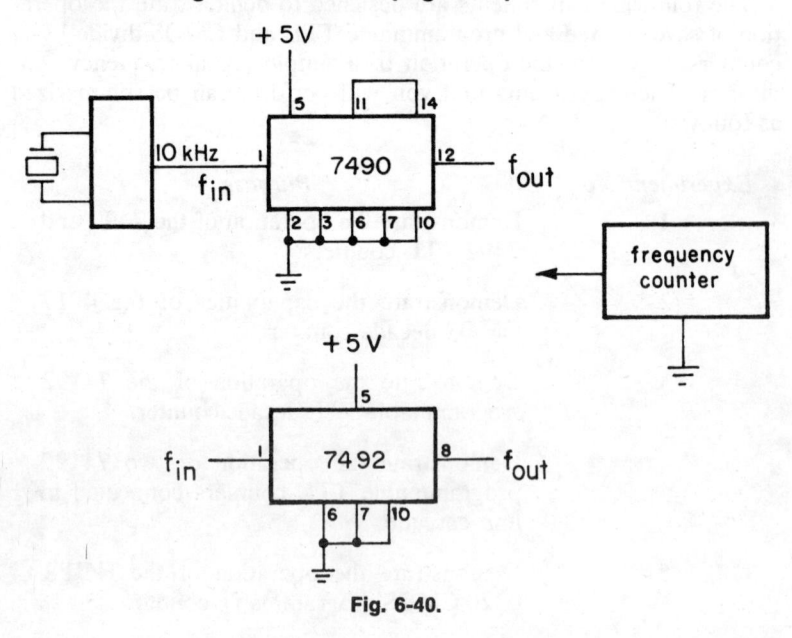

[7490 circuit: +5V to pin 5, pins 11 and 14, f_{in} 10 kHz to pin 1, f_{out} from pin 12, pins 2, 3, 6, 7, 10 to ground; feeds frequency counter]

[7492 circuit: +5V to pin 5, f_{in} to pin 1, f_{out} from pin 8, pins 6, 7, 10 to ground]

Fig. 6-40.

Step 1

Wire the two circuits shown in the schematic diagram and then apply power to the breadboard.

Step 2

Set the input frequency (using either a crystal-controlled reference such as the LR-33 Outboard or some other stable TTL square-wave source) at 10 kHz using a frequency counter. Then connect pin 1 of the 7490 integrated circuit to the frequency source and measure the output frequency at pin 12. You should find that the output frequency of the 7490 counter is $\frac{1}{10}$ the input, or 1 kHz. If your input frequency is not exactly 10 kHz, the output nevertheless should be $\frac{1}{10}$ this value, since this is a *divide-by-10* circuit.

Step 3

Next, remove the connection between pins 11 and 14 of the 7490, and connect the frequency counter to pin 11. How does the output frequency compare with the input?

The output frequency should now be $\frac{1}{5}$ that of the input, or 2.0 kHz. We now have a *divide-by-5* counter using the 7490 (see Fig. 6-15).

Step 4

Now connect the frequency source to pin 1 of the 7492 integrated circuit and the frequency counter to pin 8. What is the output frequency?

You should find that the output frequency is $\frac{1}{6}$ that of the input, or 1.666 kHz. With the 7492 device, we now have a *divide-by-6* counter circuit.

Step 5

Now connect the output of the 7492 circuit (pin 8) to the input of the 7490 circuit (pin 1) that you used in Step 3. Also connect the frequency counter to the output of the 7490 (pin 11). What is the output frequency? Why?

The output frequency should be $\frac{1}{30}$ that of the input, or 333 Hz. When cascading two or more divide-by-N counters, the modulus of the entire circuit is the produce of the modulos of the individual counters, or $6 \times 5 = 30$.

EXPERIMENT NO. 2

Purpose

The purpose of this experiment is to demonstrate the capabilities of the 4017 CMOS decade counter.

Pin Configuration of Integrated-Circuit Chip (Fig. 6-41)

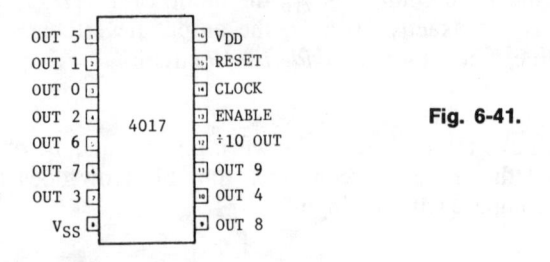

```
OUT 5 —|     |— VDD
OUT 1 —|     |— RESET
OUT 0 —|     |— CLOCK
OUT 2 —| 4017|— ENABLE      Fig. 6-41.
OUT 6 —|     |— ÷10 OUT
OUT 7 —|     |— OUT 9
OUT 3 —|     |— OUT 4
VSS   —|     |— OUT 8
```

Schematic Diagram of Circuit (Fig. 6-42)

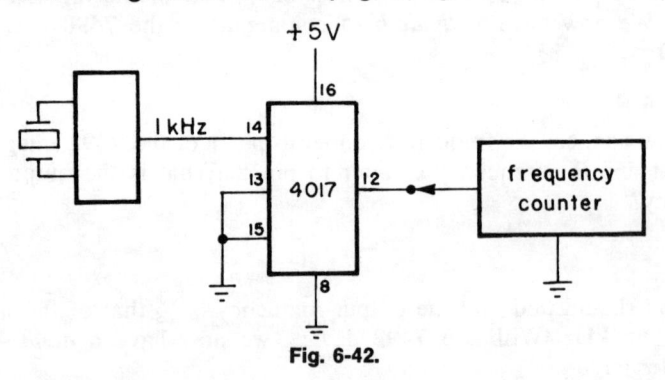

[Circuit diagram: 1 kHz source connected to pin 14 of 4017, pins 13 and 15 shown, pin 16 to +5V, pin 8 to ground, pin 12 to frequency counter.]

Fig. 6-42.

Step 1

Wire the circuit shown in the schematic diagram and apply power to the breadboard. Then connect the 10-kHz input reference frequency to pin 14, and the frequency counter to pin 12. What is the output frequency?

The output frequency should be 1.0 kHz since this is a divide-by-10 counter.

Step 2

Now sequentially connect the frequency counter to pins 1, 2, 3, 4, 5, 6, 7, 9, 10, and 11. What do you observe?

The output frequency at each of these pins is also 1.0 kHz. The difference between these outputs and the output at pin 12 is that the output at pin 12 is a symmetrical square wave, while the outputs of these pins have a positive pulse duration of 1 input cycle (10% duty cycle).

Step 3

Disconnect the reference frequency from the circuit and then disconnect the power from the breadboard. Now connect the OUT 2 pin (pin 4) to the RESET pin (pin 15) and the frequency counter to pin 3. Apply power to the breadboard and reconnect the frequency reference to pin 14. What is the output frequency?

The output frequency should be 5.0 kHz, or one-half the input frequency. Starting at 0, each decoded output sequentially goes *high* for one input cycle. After two input cycles, the output pulse at pin 4 goes *high,* and in turn resets the counter. Consequently, the counter is reset every two input cycles, giving a divide-by-2 circuit.

Step 4

Now connect pin 15 to pin 7. What is the output frequency?

The output frequency should be 3.333 kHz. By connecting the RESET input to the OUT 3 pin, the counter is reset every three input cycles, giving a divide-by-3 circuit.

Step 5

Measure the output frequency by sequentially connecting the RESET input to the decoded output pins listed in Table 6-4.

Table 6-4.

RESET Connected To:	Output Frequency	Divide-By-N
OUT 2 (pin 4)	5.000 kHz	2
OUT 3 (pin 7)	3.333 kHz	3
OUT 4 (pin 10)		
OUT 5 (pin 1)		
OUT 6 (pin 5)		
OUT 7 (pin 6)		
OUT 8 (pin 9)		
OUT 9 (pin 11)		

You should have found that it is possible to divide by any number from 2 to 10 using the 4017 counter.

EXPERIMENT NO. 3

Purpose

The purpose of this experiment is to demonstrate the operation of the 74192 programmable TTL counter.

Pin Configuration of Integrated-Circuit Chip (Fig. 6-43)

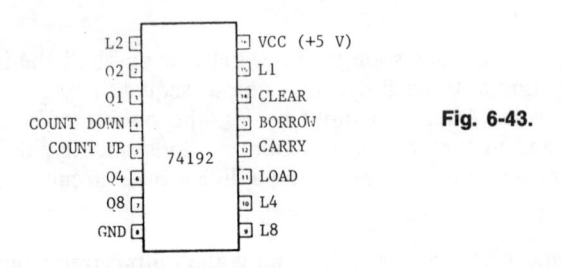

```
         L2  [1]        [16] VCC (+5 V)
         Q2  [2]        [15] L1
         Q1  [3]        [14] CLEAR
  COUNT DOWN [4]        [13] BORROW
    COUNT UP [5] 74192  [12] CARRY
         Q4  [6]        [11] LOAD
         Q8  [7]        [10] L4
         GND [8]        [9]  L8
```

Fig. 6-43.

Schematic Diagram of Circuit (Fig. 6-44)

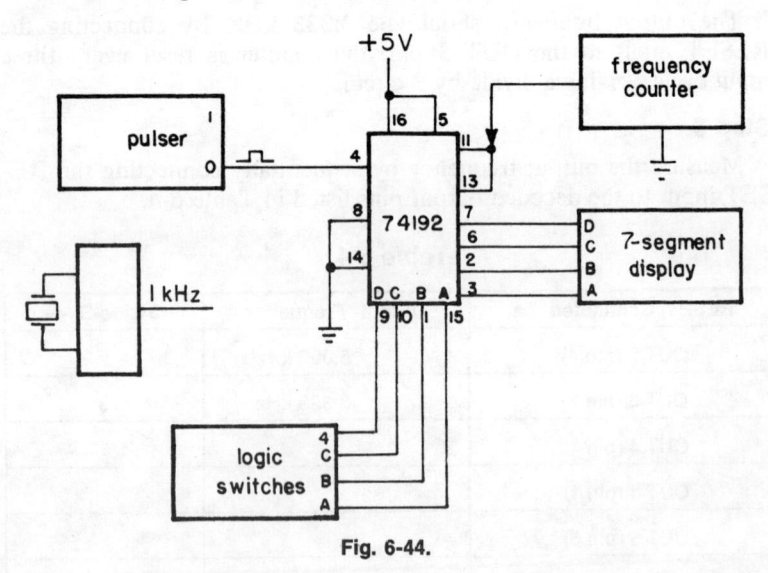

Fig. 6-44.

Step 1

Wire the circuit shown in the schematic diagram. Set the logic switches so that DCBA = 1001.

Step 2

Apply power to the breadboard. What number do you see on the LED display?

You should see the number 9, since the data inputs are initially parallel-loaded with the 4-bit bcd code 1001, which is equivalent to the decimal number 9.

Step 3

Press and release the pulser switch four times. What do you observe on the LED display?

The LED display counts backward from 9 down to 5. Continue to press and release the pulser switch five more times. What happens?

When the count reaches 0, the display then shows the number 9 when the pulser is released. This is because, when the count reaches 0, the output at pin 13 of the 74192 counter immediately goes to logic 0. The counter is then parallel-loaded with the 4-bit bcd number that is present at the data inputs (1001).

Step 4

Continue to press and release the pulser switch several times. Then set the data logic switches at DCBA = 0101. Now continue to press and release the pulser switch until the display reaches 0. When you release the pulser switch after the display indicates 0, what number do you see on the LED display?

You should now see the number 5, corresponding to the 4-bit bcd code 0101. When the counter reaches 0, the counter is then parallel-loaded with the bcd code 0101, which is the decimal number 5.

Step 5

Now set the data logic switches according to Table 6-5 and write down the number that is displayed after the count reaches 0. Your results should correspond with the 4-bit bcd number that is present at the data inputs.

Table 6-5.

D	C	B	A	LED Display
1	0	0	1	9
0	1	0	1	5
0	0	1	1	
0	0	1	0	
0	1	0	0	
0	1	1	0	
1	0	0	0	
0	0	0	1	
0	1	1	1	

Step 6

Now connect a 1-kHz signal in place of the pulser switch, and a frequency counter at pin 11. Set the logic switches at DCBA = 0001. What frequency do you observe?

Since the logic switches are set at DCBA = 0001, or the decimal number 1, the output frequency is the same as the input, or 1 kHz.

Step 7

Vary the logic switch settings according to Table 6-6, and record the output frequency for each setting.

From your measurements, you should have observed that the output frequency is that fraction of the 1-kHz input set by the decimal equivalent of the 4-bit bcd code. We then have a 1-decade programmable counter that is capable of dividing an input frequency by any integer from 1 to 9.

Step 8

What is the output frequency when you set the logic switches at DCBA = 0000?

Since division by 0 is not defined mathematically, the 74192 counter is disabled so that the output frequency is the same as the input, or 1 kHz.

EXPERIMENT NO. 4

Purpose

The purpose of this experiment is to demonstrate the operation of two 74192 programmable TTL counters connected in unit-cascade.

Pin Configuration of Integrated-Circuit Chip (Fig. 6-45)

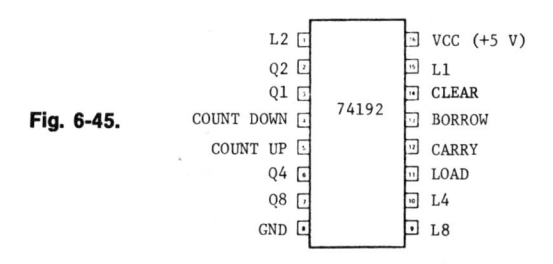

Fig. 6-45.

```
L2  [1]           [16] VCC (+5 V)
Q2  [2]           [15] L1
Q1  [3]           [14] CLEAR
COUNT DOWN [4] 74192 [13] BORROW
COUNT UP   [5]    [12] CARRY
Q4  [6]           [11] LOAD
Q8  [7]           [10] L4
GND [8]           [9]  L8
```

Schematic Diagram of Circuit (Fig. 6-46)

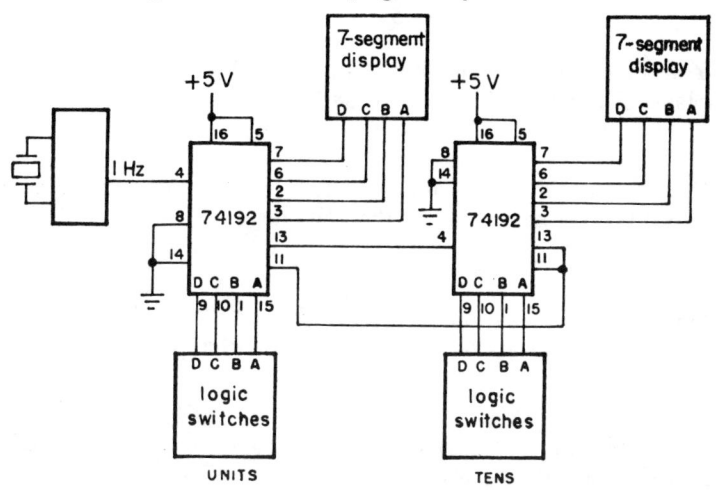

Fig. 6-46.

Step 1

Wire the circuit shown in the schematic diagram. Try to position the *tens* LED display to the left of the *units* display so that the displays will read properly. Initially set the logic switches of the units decade counter at $DCBA_1 = 0000$, and the tens decade counter at $DCBA_{10} = 0101$. In addition, temporarily disconnect the output of the 1-Hz function generator from pin 4 of the units counter.

111

Table 6-6.

D	C	B	A	N	Frequency
0	0	0	1	1	1000 Hz
1	0	0	0		
0	1	0	1		
0	0	1	1		
0	1	0	0		
1	0	0	1		
0	1	1	1		

Step 2
Apply power to the breadboard. What do you see on the LED displays?

You should see the number 8 on both the tens and units LED displays, giving an initial count of 88.

Step 3
Now connect the 1-Hz function generator to the counter circuit. What do you see happening on the LED displays?

Starting at 88, the display counts *backward*, 88, 87, 86, etc., toward 00. When the display finally reaches 00, what happens?

The display momentarily shows the number 50, and resumes counting 50, 49, 48, etc., toward 00.

Step 4
During the counting process, set the *units* logic switches at $DCBA_1 = 0100$. After the display reaches 00, at what number does the display resume counting?

The display resumes counting backward from the number 54. This is because the tens decade counter is loaded with the bcd code

112

$DCBA_{10} = 0101$ (5) and the units decade counter with $DCBA_1 = 0100$ (4).

Step 5

Now change the input frequency to 10 kHz and connect a frequency counter at pin 13 of the *tens* decade counter. What is the output frequency? Why?

If the input frequency is exactly 10.000 kHz, you should have measured an output frequency of 185 Hz. Since the modulus of the counter circuit is 54, the output frequency is $\frac{1}{54}$ times the input, or 185 Hz.

Step 6

Set the logic switches according to Table 6-7 and record the measured output frequency and determine the modulus of the circuit.

Table 6-7.

$DCBA_{10}$	$DCBA_1$	Output Frequency	Modulus
0101	0100	185 Hz	54
0000	0101		
0001	0000		
0010	0111		
0110	0011		
1001	1001		
0011	0011		
0001	0100		

Four results should show that the output frequency is that fraction of the 10-kHz input set by the logic switches. We then have a 2-decade programmable counter which, depending on the settings of the logic switches, can divide the input frequency by any integer from 1 to 99. In the following space, draw how you would cascade three 74192 counters to divide by any integer from 1 to 999. Compare your circuit with Fig. 6-26.

EXPERIMENT NO. 5
Purpose

The purpose of this experiment is to demonstrate the operation of the Hughes HCTR 0320 CMOS integrated circuit as a programmable divider.

Pin Configuration of Integrated-Circuit Chip (Fig. 6-47)

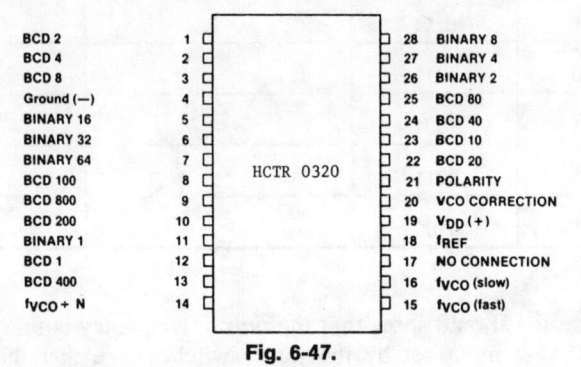

```
BCD 2        1            28   BINARY 8
BCD 4        2            27   BINARY 4
BCD 8        3            26   BINARY 2
Ground (−)   4            25   BCD 80
BINARY 16    5            24   BCD 40
BINARY 32    6            23   BCD 10
BINARY 64    7   HCTR     22   BCD 20
BCD 100      8   0320     21   POLARITY
BCD 800      9            20   VCO CORRECTION
BCD 200     10            19   V_DD (+)
BINARY 1    11            18   f_REF
BCD 1       12            17   NO CONNECTION
BCD 400     13            16   f_VCO (slow)
f_VCO ÷ N   14            15   f_VCO (fast)
```

Fig. 6-47.

Design Basics
- Binary modulus: $0 \leq N_B \leq 127$
- Decimal modulus: $0 \leq N_D \leq 999$
- Total modulus: $3 \leq N_D + N_B \leq 1023$

Schematic Diagram of Circuit (Fig. 6-48)

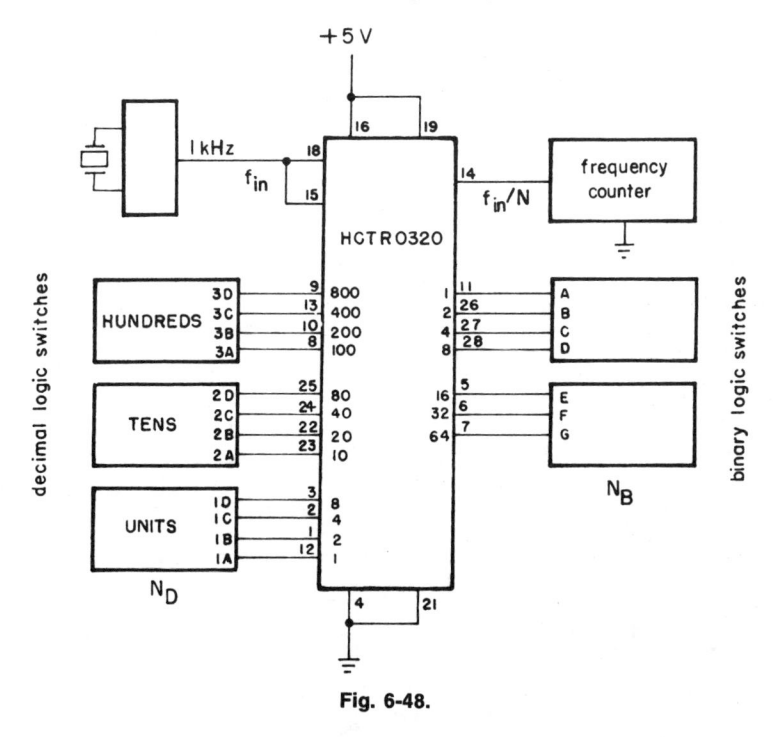

Fig. 6-48.

Step 1

The first point that should be emphasized is that the HCTR 0320 is a relatively expensive device, costing approximately $10. Therefore, you are urged to be very careful when you use it. With both the power and the frequency reference disconnected from the breadboard, wire the circuit shown in the schematic diagram. If you do not have enough logic switches, use jumper wires to connect the programming inputs either to +5 volts (logic 1) or ground (logic 0).

Step 2

Initially set the *binary* logic switches at GFEDCBA = 0000100, and all the decimal logic switches at logic 0. Then apply the power to the breadboard and connect the 1-kHz frequency reference to pins 18 and 15. Also connect the frequency counter at pin 14. What is the output frequency on the counter?

115

You should have measured an output frequency of 250 Hz since the binary logic switches are set at 0000100, which is the binary equivalent of the decimal number 4.

Step 3

Vary the binary logic switches according to Table 6-8 and record the measured output frequency and binary modulus.

Table 6-8.

G	F	E	D	C	B	A	Output Frequency	Modulus
0	0	0	0	1	0	0	250 Hz	4
0	0	0	1	0	1	0		
0	0	1	0	1	0	0		
0	1	0	0	0	1	1		
0	1	0	1	0	0	0		
1	0	0	0	0	0	1		
1	1	0	0	1	0	0		
1	1	1	1	1	0	1		

For the last 8 settings, you should have measured frequencies of 200, 83, 50, 29, 25, 15, 10, and 8 Hz, which corresponds to divisions by 5, 12, 20, 35, 40, 65, 100, and 125, respectively.

Step 4

Now change the input reference to 100 kHz. Set the *binary* logic switches at GFEDCBA = 0000111 (the decimal number 7). Also set the *units decimal* logic switches at the 4-bit code DCBA = 0011 (the decimal number 3); the *tens decimal* logic switches at DCBA = 0111 (the decimal number 7); and the *hundreds decimal* logic switches at DCBA = 0000 (the decimal number 0). Consequently, the total decimal modulus is 073. To find the *total modulus of the network, N*, we add the binary modulus to the decimal modulus. For the logic switch settings in this step, what is the programmed modulus?

The modulus is 80. Since the binary logic switches are set at 7, and the decimal logic switches are set at 73, the total modulus is 7 + 73, or 80. Now vary the binary and decimal logic switches according to Table 6-9 and record the measured output frequencies. Note that the modulos are given as a decimal number. By this time you should be able to set the switches to the corresponding logic 0s and 1s.

Table 6-9.

Binary Modulus	Decimal Modulus Hundreds	Decimal Modulus Tens	Decimal Modulus Units	Total Modulus	Output Frequency
7	0	7	3	80	
0	0	0	4	4	
1	0	0	4	5	
1	0	0	9	10	
9	0	1	1	20	
2	2	4	8	250	
4	4	9	6	500	
15	4	9	6	511	

You should have measured output frequencies that were $1/80$, $1/4$, $1/5$, $1/10$, $1/20$, $1/250$, $1/500$, and $1/511$ times the input frequency. If the input reference frequency was exactly 100 kHz, the output frequencies will be 1250 Hz, 25.0 kHz, 20.0 kHz, 10.0 kHz, 5000 Hz, 400 Hz, 200 Hz, and 196 Hz, respectively.

EXPERIMENT NO. 6

Purpose

The purpose of this experiment is to demonstrate the operation of a simple 3-decade frequency synthesizer using the MC4024, MC4044, and 74192 integrated circuits.

Pin Configuration of Integrated-Circuit Chips (Fig. 6-49)

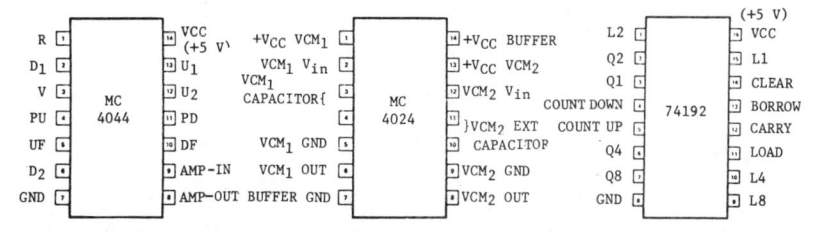

Fig. 6-49.

117

Schematic Diagram of Circuit (Fig. 6-50)

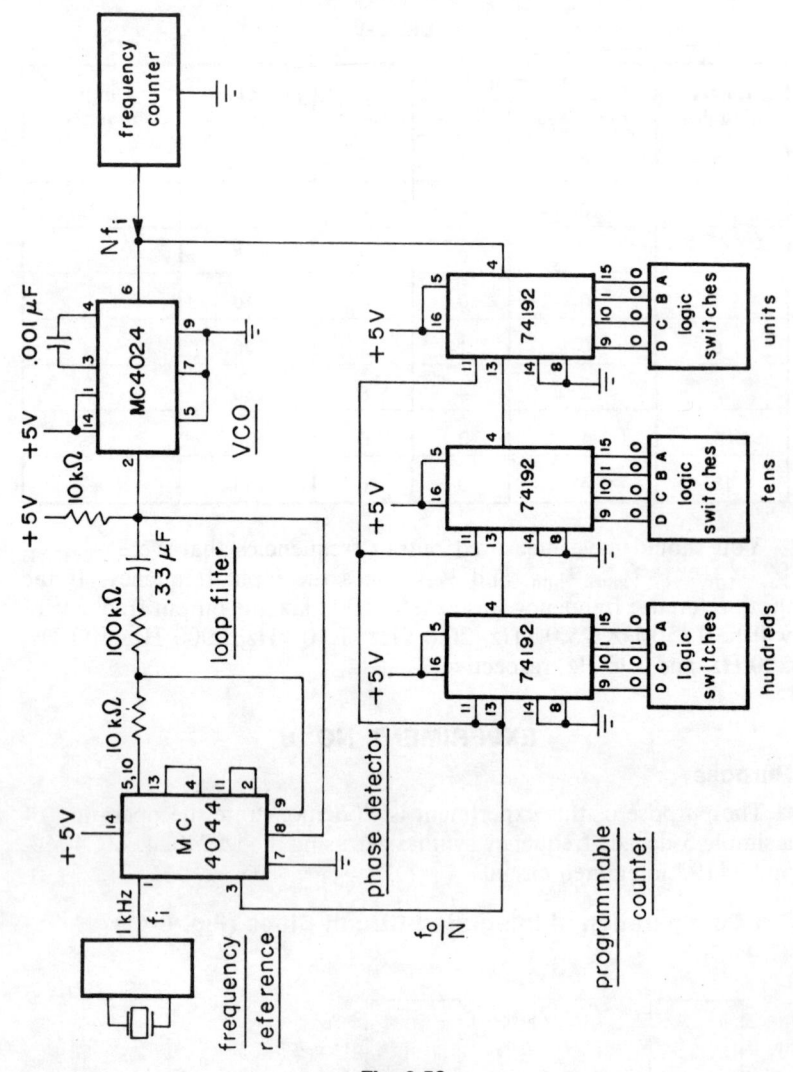

Fig. 6-50.

Step 1

Wire the circuit shown in the schematic diagram. Because of the large number of components, logic switches, etc., you will probably need two SK-10 breadboarding sockets to wire this circuit. In addition, *you must use a stable, crystal-controlled 1.000-kHz frequency reference.*

Step 2

Apply power to the breadboard. What frequency do you measure with the frequency counter?

If the input frequency reference is exactly 1.000 kHz, you should measure an output frequency of the synthesizer that is exactly 200.000 kHz. The three programmable counters are set to divide by 200. Consequently, the output frequency is 200 times the input reference, or 200 kHz. If it is not exactly this value, within 1 kHz, your input reference is off a little. To determine the actual input frequency, divide the measured output frequency of the synthesizer by 200, and record your result:

$$f_{REF} = \underline{\qquad} Hz$$

If your output frequency is significantly different than 200 kHz, perform the next step. Otherwise, go directly to Step 4.

Step 3

If you do not measure an output frequency close to 200 kHz, disconnect the 1.0-kHz input frequency reference from the circuit and measure the synthesizer's output frequency. With no input reference signal, the output frequency should be about 145 kHz, which is the *vco free-running frequency.* If it is somewhat *lower* than 145 kHz, substitute a different 0.001-μF capacitor between pins 3 and 4 of the MC4024 integrated circuit until you obtain a free-running frequency around 145 kHz. Reconnect the 1-kHz input frequency reference to pin 1 of the MC4044 phase detector. You should now measure an output frequency close to 200 kHz.

If the preceding does not work, check the 4-bit bcd code at the data inputs to each of the three 74192 programmable counters (pins 15, 1, 10, and 9). The *hundreds* decade counter should have a bcd input of DCBA = 0010 (the decimal number 2), while the inputs to the other two 74192 counters should all be at logic 0. If the logic switch settings are correct and the output frequency is still not 200 kHz, then carefully check your wiring against the schematic diagram.

Step 4

Now change the logic switch settings of the *tens* decade counter to DCBA = 0101. What is the output frequency now? Why?

The measured output frequency should be 250 times the input reference frequency you determined in Step 2. The modulus of the cascaded programmable counters is 250.

Step 5

Now change the logic switch settings of the *units* decade to DCBA = 1001. What is the output frequency now? Why?

The measured output frequency should be 259 times the input reference frequency you determined in Step 2.

Step 6

Now change the logic switch settings of the *hundreds* decade to DCBA = 0011. What is the output frequency?

The measured output frequency should be 359 times the input reference frequency you determined in Step 2.

Step 7

Vary the logic switch setting for any number between 200 and 400. You should measure an output frequency that is this number times the input reference frequency you determined in Step 2.

Step 8

Now set the logic switch settings at 100 (the *hundreds* decade at DCBA = 0001, the *tens* and *units* decades at DCBA = 0000). What is the synthesizer's output frequency?

The output frequency should be about 145 kHz, which is the *vco free-running frequency* (see Step 3). For the MC4024 integrated circuit vco, the output frequency of the synthesizer cannot go below the free-running frequency. Consequently, the vco free-running frequency must be less than the lowest output frequency expected from the synthesizer. Now place another 0.001-μF capacitor between pins 3 and 4 of the MC4024 integrated circuit device, so that there are now two 0.001-μF capacitors *in parallel*. You should now measure an output frequency that is 100 times the input frequency you determined in Step 2.

Step 9

Disconnect the 1.0-kHz input reference from the circuit and note the synthesizer's output frequency. It will probably be around 70 kHz. For the moment, this vco frequency is not too important. The only restriction is that *it must be less than the product of the modulus of the programmable counter times the input reference frequency* (i.e., 100 times 1.0 kHz, or 100 kHz).

CHAPTER 7

Monolithic Integrated Circuits and Applications

INTRODUCTION

This chapter describes several popular monolithic phase-locked-loop devices. These are integrated circuits that contain a phase detector, vco, plus several specialized functions within a single package. All that is required is several external resistors and capacitors to set the vco frequency and the loop filter. Routine applications such as am and fm detectors, fsk decoders, and prescalers for frequency counters are ideally suited for these devices. For frequency synthesis, it is necessary to add an external divide-by-N counter between the vco and the phase detector. Only brief descriptions of some of the monolithic devices are given, as the major operating characteristics and necessary design information are fully described in the data sheets in Appendix B.

OBJECTIVES

At the completion of this chapter you will be able to do the following:

- Be familiar with several of the 560 series of monolithic phase-locked-loop devices.
- Be familiar with the operation of the 4046 CMOS phase-locked loop.
- Perform an experiment that uses a 565 phase-locked loop as a frequency-shift-keying (fsk) demodulator.

- Perform an experiment that uses a 567 phase-locked-loop tone decoder.
- Perform an experiment that uses a 4046 phase-locked loop as a frequency synthesizer, or multiplying prescaler.
- Describe several useful applications using monolithic phase-locked-loop devices.

THE 560 SERIES

The 560 series of monolithic phase-locked-loop devices was first introduced by Signetics Corporation. This series includes the 560B, 561B, 562, 564, 565, and 567. Not all of these devices will be discussed in this chapter. However, the data sheets for all six are given in Appendix B. The 560 series devices are commonly referred to as *analog* phase-locked loops. Up to now, the phase-locked loops discussed in this book have been *digital*. The basic difference between analog and digital phase-locked loops is the type of phase detector used. The exclusive-OR and edge-triggered detectors described in Chapter 3 are digital. Almost all analog phase-locked-loop systems use a *double-balanced mixer*. Although the digital detectors were fully explained earlier, we will not discuss the analog detector. It is not too important to know how it works, but rather to remember that the output of such a detector is an average voltage that is proportional to the phase difference between its two inputs.

560B Phase-Locked Loop

The 560B (Fig. 7-1) is the most fundamental of the series. It contains a phase detector, amplifier, and vco in a 16-pin package. When locked onto an input signal, two useful outputs are provided. First, an output voltage that is proportional to the frequency of the incoming signal is available as the DEMODULATED FM OUTPUT at pin 9. The second output is the square-wave output signal of the vco. The value of the external capacitor, connected between pins 2 and 3, required to set the vco free-running frequency (f_o) is given by

$$C_o(pF) = \frac{300}{f_o} \qquad \text{(Eq. 7-1)}$$

The 560B is used primarily as an fm demodulator, using the basic circuit of Fig. 7-2. The vco is tuned by adjusting the external capacitor (C_o) to the center frequency of the fm signal. For most applications, the pair of loop filter capacitors (C_1) can be determined from the approximate relation:

$$C_1(\mu F) \simeq \frac{13.3}{f_{3\,dB}} \qquad \text{(Eq. 7-2)}$$

where $f_{3\,dB}$ is the desired bandwidth of the demodulated information.

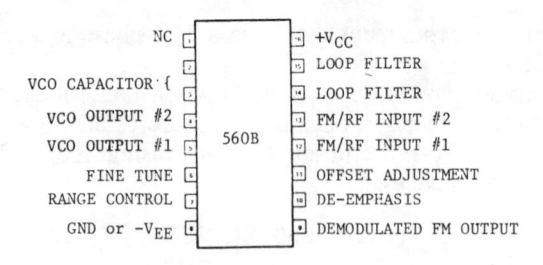

```
        NC  ⃞        ⃞  +V_CC
VCO CAPACITOR {⃞     ⃞  LOOP FILTER
                ⃞    ⃞  LOOP FILTER
    VCO OUTPUT #2 ⃞  ⃞  FM/RF INPUT #2
    VCO OUTPUT #1 ⃞ 560B ⃞ FM/RF INPUT #1
        FINE TUNE ⃞  ⃞  OFFSET ADJUSTMENT
    RANGE CONTROL ⃞  ⃞  DE-EMPHASIS
       GND or -V_EE ⃞ ⃞  DEMODULATED FM OUTPUT
```

Fig. 7-1. Pin diagram of the 560B phase-locked loop.

The DEMODULATED FM OUTPUT at pin 9 is an output voltage that is a function of the frequency deviation of the input signal. For a ±1% deviation, the output is approximately 0.3 volt peak-to-peak (0.11 volt rms). As an example, a standard 10.7-MHz if (intermediate frequency) circuit has a deviation of approximately ±75 kHz. The percent deviation is then:

$$\% \text{ deviation} = \frac{\pm 75 \text{ kHz}}{10.7 \text{ MHz}} \times 100$$
$$= \pm 0.7\%$$

Consequently, the output voltage swing is

$$V_o = \frac{0.3 \text{ volt p-p}}{\pm 1\% \text{ deviation}} \times (\pm 0.7\%)$$
$$= 0.21 \text{ volt peak-to-peak } (0.075 \text{ volt rms})$$

$R_1 = R_2 = 50\ \Omega$

Fig. 7-2. An fm demodulator using the 560B phase-locked loop.

For additional design information, consult the 560B data sheet in Appendix B.

561B Phase-Locked Loop

The 561B, shown in Fig. 7-3, is identical to the 560B, except that it includes an additional phase detector, which allows the device to be used as a synchronous am detector (like the homodyne receiver). As with the 560B, the 561B can be used for fm demodulation.

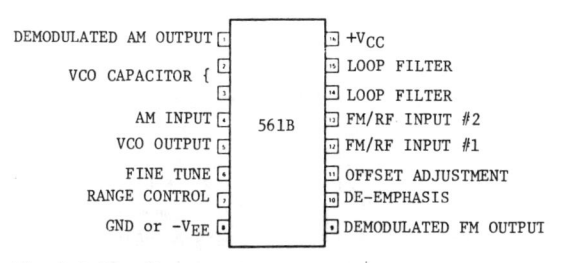

```
DEMODULATED AM OUTPUT [1]        [16] +V_CC
         VCO CAPACITOR { [2]     [15] LOOP FILTER
                         [3]     [14] LOOP FILTER
              AM INPUT [4] 561B  [13] FM/RF INPUT #2
             VCO OUTPUT [5]      [12] FM/RF INPUT #1
              FINE TUNE [6]      [11] OFFSET ADJUSTMENT
          RANGE CONTROL [7]      [10] DE-EMPHASIS
            GND or -V_EE [8]     [9] DEMODULATED FM OUTPUT
```

Fig. 7-3. Pin diagram of the 561B phase-locked loop.

A simple am broadcast-band (550 to 1600 kHz) receiver using the 561B is shown in Fig. 7-4. Unlike other receivers, there is no capacitor-inductor tuning circuit! The tuning capacitor connected between pins 2 and 3 is selected to make the vco oscillate at the frequency to be received. For broadcast-band operation, a 365-pF variable capacitor is required.

In operation, this receiver circuit requires an outside antenna and a good ground. In addition, sufficient signal must be present at the input of the phase-locked loop. Otherwise a "swishing" sound may result, which is due to the frequency difference between the incoming carrier and the vco when the loop lock is unstable. Another drawback

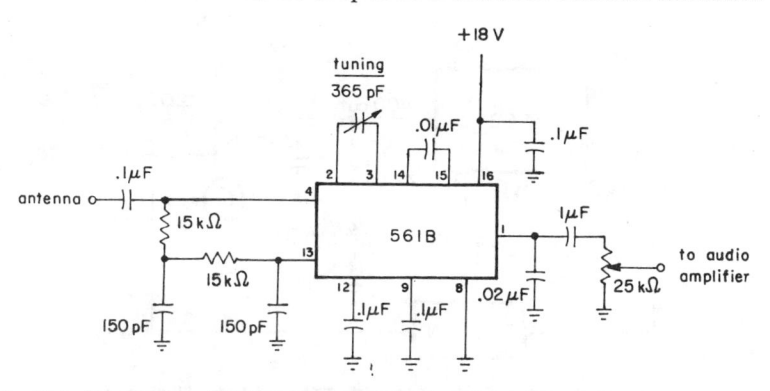

Fig. 7-4. An am broadcast-band receiver using the 561B phase-locked loop.

125

of this simple circuit is the "hand-capacitance" effect, which is due to the nongrounded capacitive tuning of the vco. One remedy for this problem is to use a vernier dial and an insulated shaft on the tuning capacitor. Operation of this circuit may be improved by using an untuned broadband amplifier ahead of the receiver to increase the sensitivity. However, care should be used to ensure that the input voltage to the 561B device does not exceed 0.5 volt rms. An easy-to-read article describing such a phase-locked-loop receiver appears on Page 58 in the October, 1971 issue of *Ham Radio*.

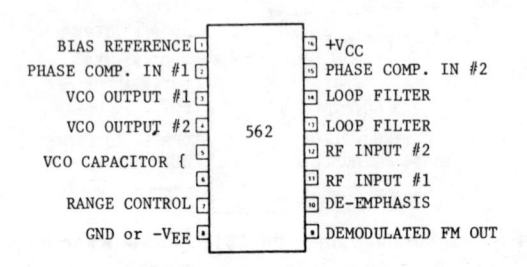

```
       BIAS REFERENCE [1]          [16] +V_CC
       PHASE COMP. IN #1 [2]       [15] PHASE COMP. IN #2
           VCO OUTPUT #1 [3]       [14] LOOP FILTER
           VCO OUTPUT #2 [4]  562  [13] LOOP FILTER
           VCO CAPACITOR { [5]     [12] RF INPUT #2
                           [6]     [11] RF INPUT #1
           RANGE CONTROL [7]       [10] DE-EMPHASIS
              GND or -V_EE [8]     [9]  DEMODULATED FM OUT
```

Fig. 7-5. Pin diagram of the 562 phase-locked loop.

562 Phase-Locked Loop

The 562 monolithic phase-locked-loop device, shown in Fig. 7-5, is basically the same as the 560 B, except that the internal connection between the vco output and the phase comparator is broken, providing two external connections. This feature allows for a TTL divide-by-N counter to be placed in the feedback path for frequency synthesis, as shown in Fig. 7-6.

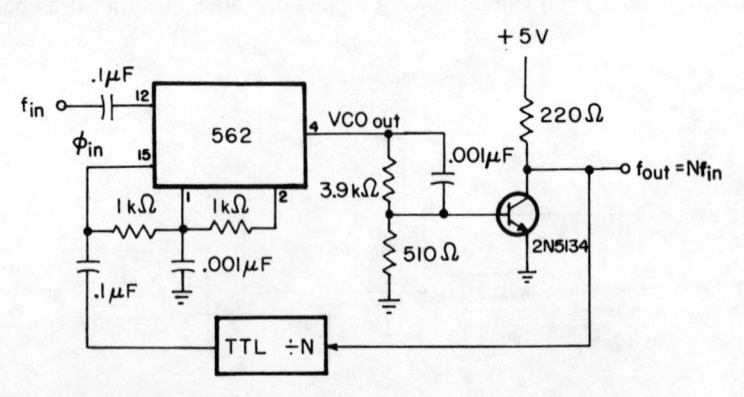

Fig. 7-6. Interfacing the 562 with TTL divide-by-N counters.

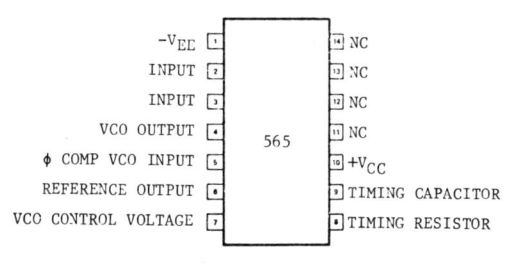

```
       -V_EE  [1]           [14] NC
       INPUT  [2]           [13] NC
       INPUT  [3]           [12] NC
  VCO OUTPUT  [4]    565    [11] NC
φ COMP VCO INPUT [5]        [10] +V_CC
REFERENCE OUTPUT [6]        [9] TIMING CAPACITOR
VCO CONTROL VOLTAGE [7]     [8] TIMING RESISTOR
```

Fig. 7-7. Pin diagram of the 565 phase-locked loop.

565 Phase-Locked Loop

The 565 phase-locked-loop device, shown in Fig. 7-7, is perhaps the most popular of the 560 series. It is a general-purpose device, similar to the 562. Whereas the 560B, 561B, and 562 devices are used for frequencies up to 30 MHz, the 565 is limited to frequencies below 500 kHz.

A generalized circuit is shown in Fig. 7-8. The vco free-running frequency is approximately determined from

$$f_o = \frac{1.2}{4R_1 C_1} \qquad \text{(Eq. 7-3)}$$

Capacitor C_1 can be any value, but resistor R_1 should be between 2 kΩ and 20 kΩ. A small capacitor, typically 0.001 μF, is usually placed between pins 7 and 8 to eliminate spontaneous oscillations. A simple first-order loop filter is formed by capacitor C_2 and an internal resistance of approximately 3.6 kΩ.

One popular application for the 565 phase-locked loop is a *frequency-shift-keying* (fsk) demodulator. Frequency-shift keying refers to data transmission by means of a carrier frequency which is shifted

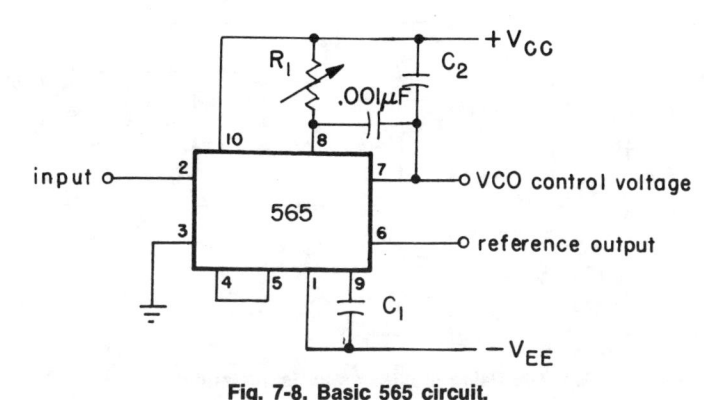

Fig. 7-8. Basic 565 circuit.

between two preset frequencies, and is widely used with teletypewriter (tty) systems, both in the computer peripheral and radio communications field. Over the years, several standards have been used to set the *mark* and *space* frequencies, which correspond to the logic 1 and logic 0 states of the binary data signal. Several of these frequency pairs are listed in Table 7-1.

Table 7-1.

Mark	Space
1070 Hz	1270 Hz
2125 Hz	2975 Hz
2025 Hz	2225 Hz

The frequency difference between the mark and space frequencies is called the *frequency shift*. For the 1070–1270-Hz pair, the frequency shift is 200 Hz. When transmitting teletypewriter information using a modulator-demodulator system, commonly referred to as a *modem,* this frequency pair usually represents the *originate* signal, while the 2025–2225-Hz pair represents the *answer* signal. By law, radio teletypewriter (rtty) must have a frequency shift of less than 900 Hz. Radio amateur operators many years ago adopted the 2125–2975-Hz (850-Hz shift) standard. A simple circuit using the 565 device as a 1070–1270-Hz, fsk demodulator is shown in Fig. 7-9. As the fsk signal appears at the input, the 565 phase-locked loop locks onto the input frequency and tracks it between the mark and space frequencies with a corresponding dc voltage shift at the output. The free-running frequency of the vco is adjusted with the 5-kΩ potentiometer

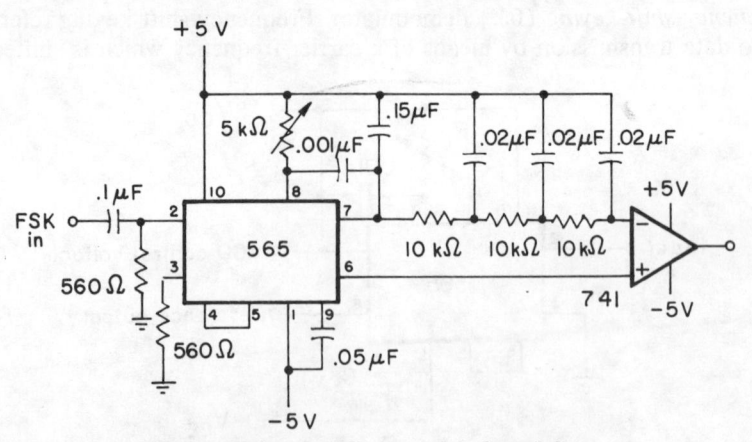

Fig. 7-9. Using the 565 as an fsk demodulator.

so as to give a slightly positive output voltage with an input frequency of 1070 Hz.

Another popular application for the 565 device is the SCA (Subsidiary Communications Authorization) decoder circuit of Fig. 7-10. Some fm radio stations are authorized by the FCC to broadcast uninterrupted background music for commercial use in stores, factories, offices, etc. To accomplish this, SCA programming is transmitted by a 67-kHz subcarrier, so as not to interfere with the fm station's main channel program, which can be either in mono or stereo. In addition, the level of this subcarrier is only 10% of the amplitude of the combined signal. The input is connected to the fm tuner's detector stage

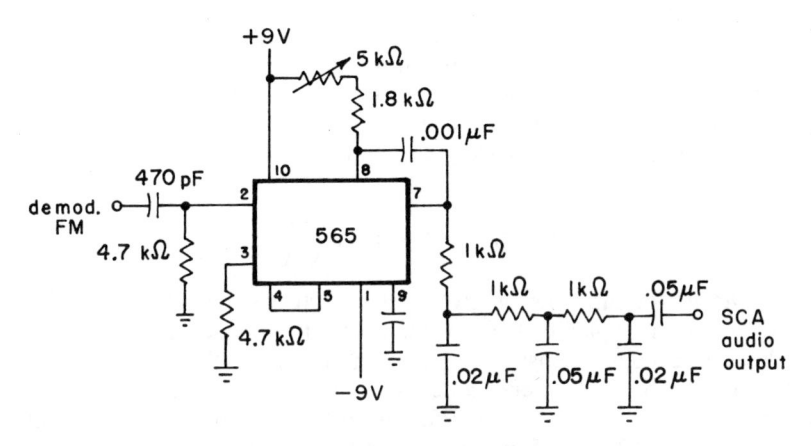

Fig. 7-10. Using the 565 as an SCA decoder.

(before the de-emphasis network), while the output of the decoder can be connected to any audio amplifier, since the SCA audio response is limited to a maximum of 7 kHz. If you have one of the older fm tuners with an *MPX* output, or a modern tuner with an *fm detector, 4-channel, or fm quadraphonic* output jack, connect the decoder's input to it, as this is the output of the fm detector before the de-emphasis network. Once connected, the decoder is tuned to 67 kHz with the 5-kΩ potentiometer.

567 Phase-Locked Loop/Tone Decoder

The 567 device (Fig. 7-11) is a phase-locked-loop system designed specifically to respond to a given tone of constant frequency within its bandwidth. Similar to the 561, the 567 has, in addition, a power output stage capable of sourcing 100 mA. Its frequency range, however, is similar to the 565, which is limited to 500 kHz.

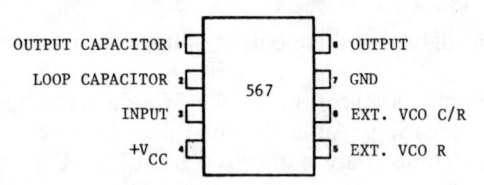

Fig. 7-11. Pin diagram of the 567 phase-locked loop/tone decoder.

Fig. 7-12 shows the basic connections for the 567 tone decoder. The free-running, or center frequency (f_o) of the vco is set by R_1 and C_1, so that

$$f_o \cong \frac{1.10}{R_1 C_1} \qquad \text{(Eq. 7-4)}$$

where R_1 should be between 2 kΩ and 20 kΩ. The value for C_2 is best found from the "Bandwidth versus Input Signal Amplitude" graph presented in the 567 data sheet in Appendix B. The value for C_3 is not critical, but should be at least twice that of C_2.

For input signal levels (V_i) less than 200 mV rms, the bandwidth of the loop is found from

$$\text{BW (\% of } f_o) \cong 1070 \left(\frac{V_i}{f_o C_2}\right)^{1/2} \qquad \text{(Eq. 7-5)}$$

However, for input levels greater than 200 mV rms, the bandwidth of the 567 is typically 14% of the center frequency. In addition, the decoder becomes sensitive to input frequencies that are odd subharmonics of the center frequency, so that the loop may lock onto frequencies $f_o/3$, $f_o/5$, etc. Furthermore, the loop may lock onto signals near $(2n + 1)f_o$ where n = 1, 2, 3, etc. If such signals are an-

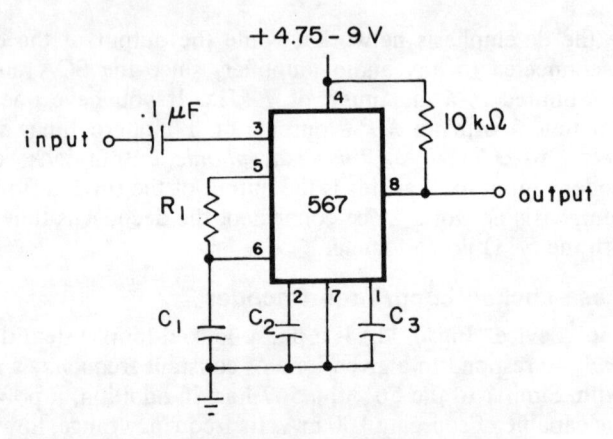

Fig. 7-12. Basic 567 tone-decoder circuit.

130

ticipated, they should be attenuated *before* reaching the input of the 567. When the loop is locked, the output at pin 8 is at logic 0.

Example

Using the basic tone-decoder circuit of Fig. 7-12, determine the values for R_1, C_1, C_2, and C_3 to decode a 100-mV, 700-Hz input signal. In addition, the bandwidth should be approximately 12% of the center frequency.

First, to determine R_1 and C_1, we choose a suitable value for C_1. Picking 0.1 μF, for example, R_1 is then found from Equation 7-4, so that

$$R_1 = \frac{1.10}{f_o C_1}$$
$$= \frac{1.10}{(700 \text{ Hz})(0.1 \text{ μF})}$$
$$= 15.7 \text{ k}\Omega$$

for which we can use a 15-kΩ, 5% resistor or a 15.8-kΩ, 1% resistor. Capacitor C_2 is then determined by rearranging Equation 7-5:

$$C_2 = \frac{V_1}{f_o} \left(\frac{1070}{BW}\right)^2 \text{ (in μF)}$$
$$= \frac{(0.10 \text{ V})}{(700 \text{ Hz})} \left(\frac{1070}{12\%}\right)^2$$
$$= 1.14 \text{ μF}$$

for which we can use a 1-μF capacitor. Since C_3 should be at least twice that of C_2, we can select C_3 to be 3.3 μF. The completed circuit is shown in Fig. 7-13.

A popular application for the 567 decoder is the decoding of *Touch-Tone®* signals. Touch-Tone® information is coded in tone-pairs, using two of seven possible tones for numbers 0 through 9, and the symbols # (pound) and * (star). The audio frequencies used are

Table 7-2. Touch-Tone® Frequencies

Low Tone Group (Hz)	High Tone Group		
	1209 Hz	1336 Hz	1477 Hz
697	1	2	3
770	4	5	6
852	7	8	9
941	*	0	#

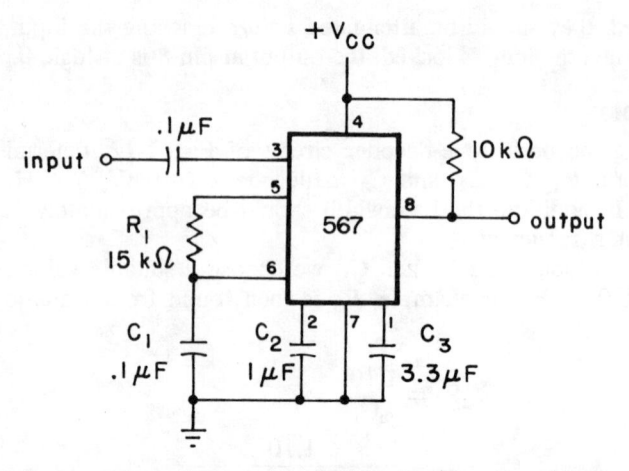

Fig. 7-13. Completed 567 tone-decoder circuit with values for decoding a 100-mV, 700-Hz input signal.

listed in Table 7-2. The basic decoder circuit for a single digit or symbol (e.g., the number 9) is shown in Fig. 7-14.

The number 9 simultaneously has a low tone of 852 Hz and a high tone of 1477 Hz. Consequently, two 567 decoders are needed. One is set for a center frequency of 852 Hz, while the other is set to 1477 Hz. When the Touch-Tone® signal corresponding to the number 9 is present at the circuit's input, the output of both decoders will be at logic 0, since both loops are now locked. The output of the NOR

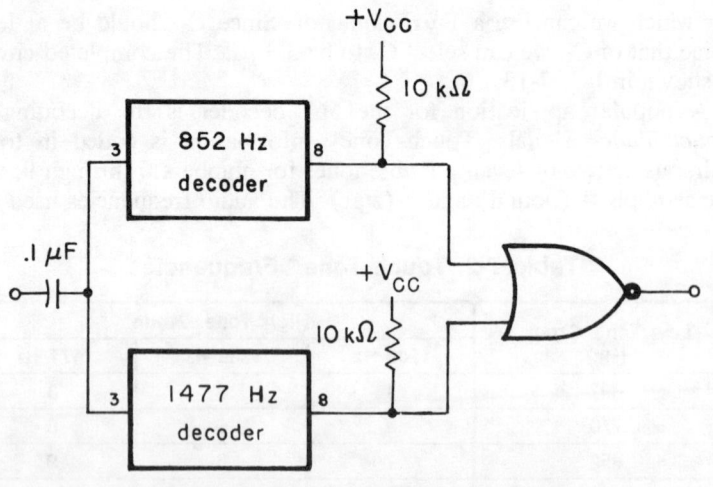

Fig. 7-14. Touch-Tone® decoder for the number 9.

gate will then be at logic 1. If only one of the two tones is present, only one of the two outputs will be at logic 0, so that the output of the NOR gate will be at logic 0. Using the basic circuit of Fig. 7-14, a complete Touch-Tone® decoder circuit, capable of decoding all possible 12 tone-pairs, is shown in Fig. 7-15.

Another novel application for the 567 tone decoder is the construction of a low-cost frequency indicator, using the circuit of Fig. 7-16. One decoder (U1) is set approximately 6% *above* the desired frequency, while the other decoder (U2) is set 6% *below*. If the input frequency is within 13% of the desired frequency, either lamp No. 1 or lamp No. 2 will come on. If both are on, then the input frequency is within 1% of the desired frequency.

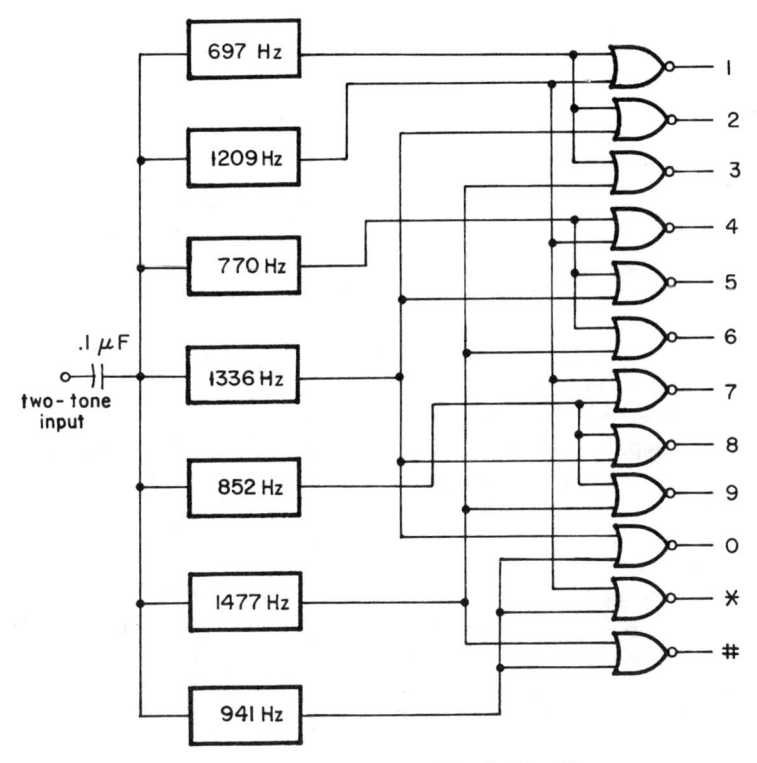

Fig. 7-15. A 12-digit, Touch-Tone® decoder.

THE 4046 CMOS PHASE-LOCKED LOOP

The 4046 CMOS monolithic phase-locked loop is currently made by RCA and comes in a 16-pin DIP case, as shown in Fig. 7-17. One of the major differences between the 4046 and the 560 series of mono-

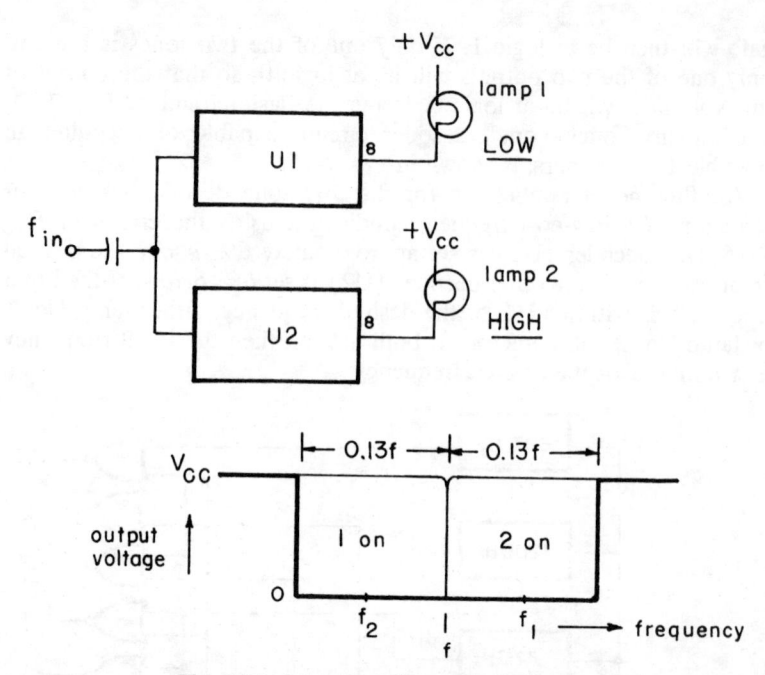

Fig. 7-16. A low-cost frequency indicator.

lithic devices is that the 4046 phase-detector system is digital rather than analog. Furthermore, the 4046 contains *two different types* of phase detectors. Referring to the block diagram of Fig. 7-18, the inputs to phase detectors I and II are connected in parallel. The outputs, however, are brought out separately. Phase detector I (sometimes called the *low-noise detector*) is a simple exclusive-OR type. Consequently, both the input and vco signals must be 50% duty cycle square waves.

Phase detector II (sometimes called the *wideband detector*) is an edge-triggered digital type, which triggers on the positive leading

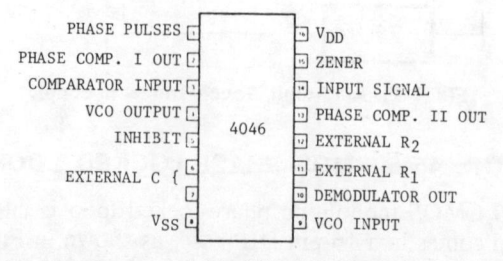

Fig. 7-17. Pin diagram of the 4046 CMOS phase-locked loop.

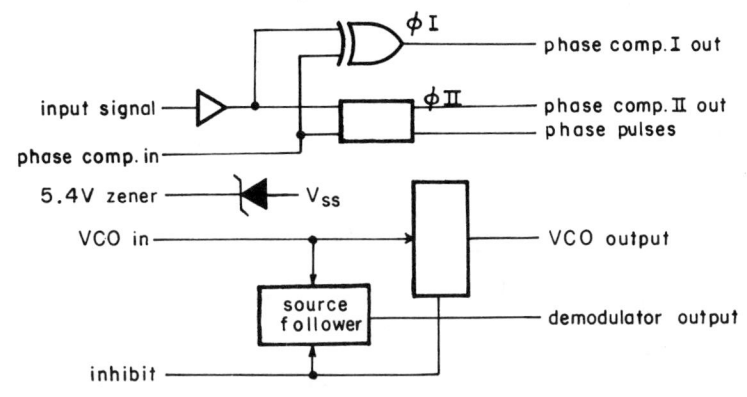

Fig. 7-18. Block diagram of the 4046 CMOS phase-locked loop.

edges of the inputs. If the input signal, which can be a pulse train having *any* duty cycle, is lower than the vco frequency, the output is at logic 0 (V_{SS}, or ground). On the other hand, if the input frequency is higher than the vco frequency, the output is at logic 1 ($+V_{DD}$). If both frequencies are the same, the output of phase detector II is a pulse whose width is proportional to the phase difference. As shown in Fig. 7-19, this output pulse is positive when the vco signal lags the input, and negative when the vco leads the input. One advantage of phase detector II over phase detector I is that the former is insensitive to harmonics, while the exclusive-OR type may lock onto harmonic multiples of the input frequency.

Also shown in the block diagram is an uncommitted 5.4-volt zener diode, which can be used if regulation of the power supply voltage is

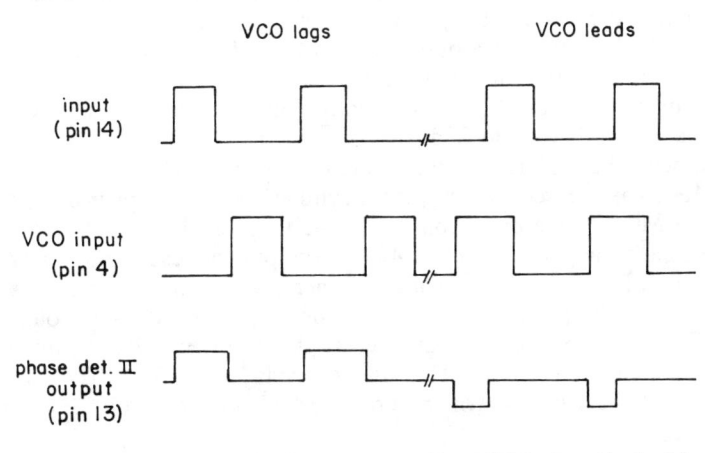

Fig. 7-19. Input/output waveforms for the 4046 CMOS phase-locked loop.

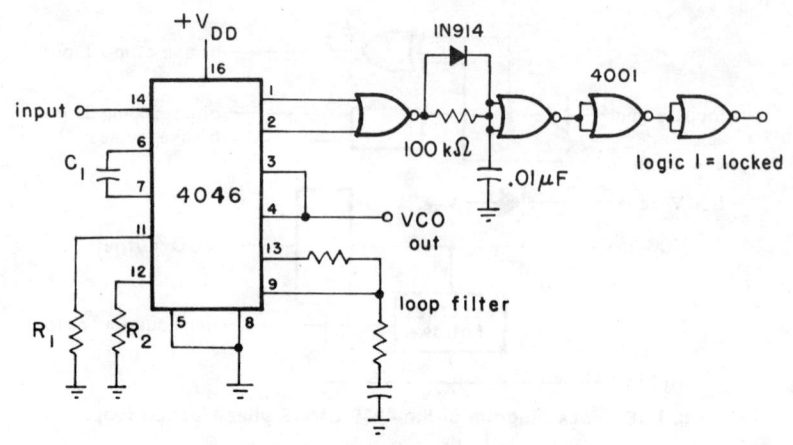

Fig. 7-20. A 4046 lock-detector circuit.

necessary. The frequency of the vco is a minimum (f_{min}) when the input control voltage is zero, and increases linearly to a maximum (f_{max}) when the control voltage equals $+V_{DD}$. Typically the maximum vco frequency can be 700 kHz when V_{DD} equals +5 volts, and 1.9 MHz when V_{DD} is +15 volts. The frequency range ($f_{max} - f_{min}$) of the vco is set by the external components R_1 and C_1, while the minimum vco frequency is controlled by R_2 and C1*. Resistors R_1 and R_2 should be between 10 kΩ and 1 MΩ, while C_1 should be greater than 100 pF for $V_{DD} \geqq 5$ volts, or greater than 50 pF for $V_{DD} \geqq 10$ volts. If, for some applications, it is not necessary to have the vco on at certain times, the vco can be turned off by tying the INHIBIT input (pin 5) to $+V_{DD}$, which will also minimize power consumption. Otherwise, the INHIBIT input is connected to V_{SS}.

The PHASE PULSES output (pin 1) of phase detector II can be NORed with the output of phase detector I (pin 2) to form a lock detector, as shown in Fig. 7-20. The output of the last 4001 NOR gate will be a logic 1 when locked.

As with the majority of the monolithic phase-locked-loop devices, the 4046 can be used as a frequency synthesizer or multiplier by inserting a CMOS divide-by-N counter in the loop feedback path between the vco and the phase detector. As an example, the resolution of a typical digital frequency counter may be increased from ±1 Hz to ±0.01 Hz for low-frequency signals by using a pair of decade counters (74C90 or 4018, etc.) in cascade, as shown in Fig. 7-21. Thus, an input frequency of 52.83 Hz will be displayed as 5283 Hz on the counter. Otherwise, all you would observe is either 52 or 53 Hz. By

---

*Vco design information is given in the 4046 data sheet in Appendix B.

making the modulus of the divide-by-N counter equal to 60, the same technique can be used to display the input frequency in terms of *cycles per minute*. This is extremely useful when measuring the frequency of physiological exents such as respiration (breaths per minute), or heart rate (beats per minute).

AN INTRODUCTION TO THE EXPERIMENTS

The following experiments are designed to illustrate the opration of several types of monolithic phase-locked-loop devices with several applications. The experiments that you will perform can be summarized as follows:

Experiment No.   *Purpose*

1   Demonstrates the operation of the 565 phase-locked loop as a frequency-shift-keying (fsk) demodulator.

2   Demonstrates the operation of the 567 phase-locked-loop tone decoder.

3   Demonstrates the operation of the 4046 CMOS phase-locked loop.

4   Demonstrates the function of a "loss-of-lock" indicator with the 4046 CMOS phase-locked loop.

5   Demonstrates the operation of a CMOS frequency synthesizer that can be used as a multiplying prescaler for frequency counters.

EXPERIMENT NO. 1

Purpose

The purpose of this experiment is to demonstrate the operation of the 565 phase-locked loop as a frequency-shift-keying (fsk) demodulator. In addition, two 555 timers are used as a simple fsk generator.

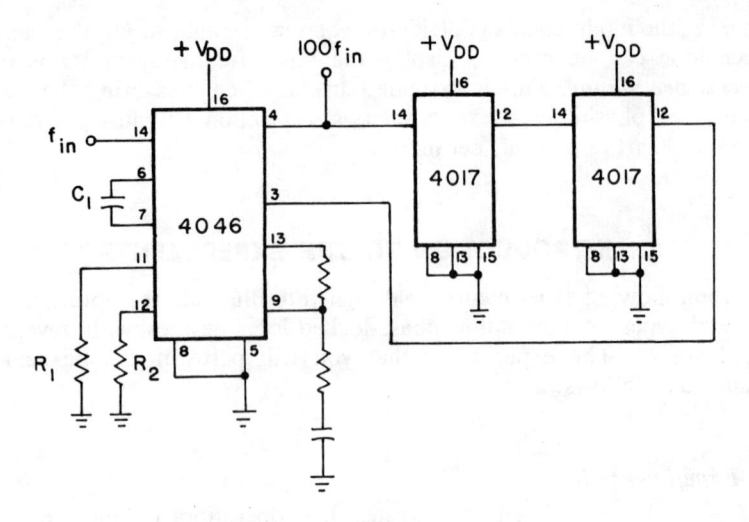

Fig. 7-21. A ×100 frequency multiplier.

Pin Configuration of Integrated-Circuit Chips (Fig. 7-22)

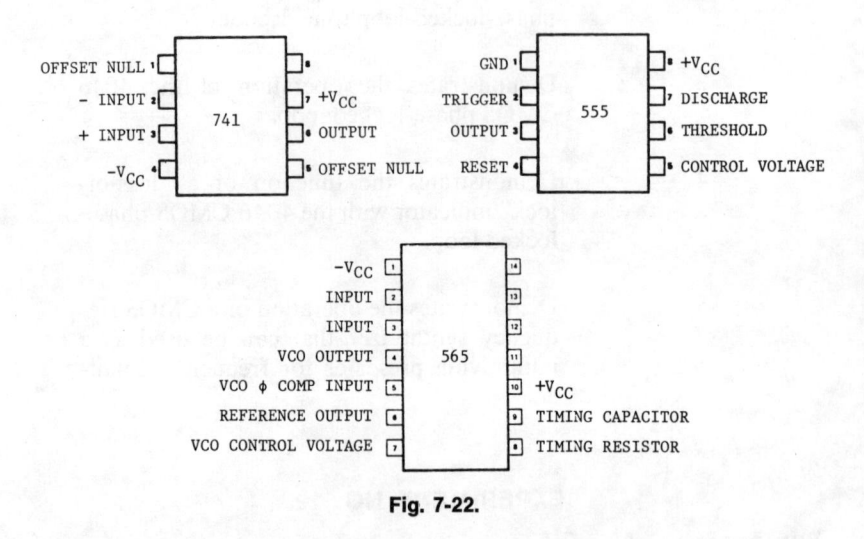

Fig. 7-22.

Schematic Diagram of Circuits (Fig. 7-23)

The schematic diagram of the two circuits for this experiment is shown in Fig. 7-23. Circuit A is the fsk generator and circuit B is the fsk demodulator.

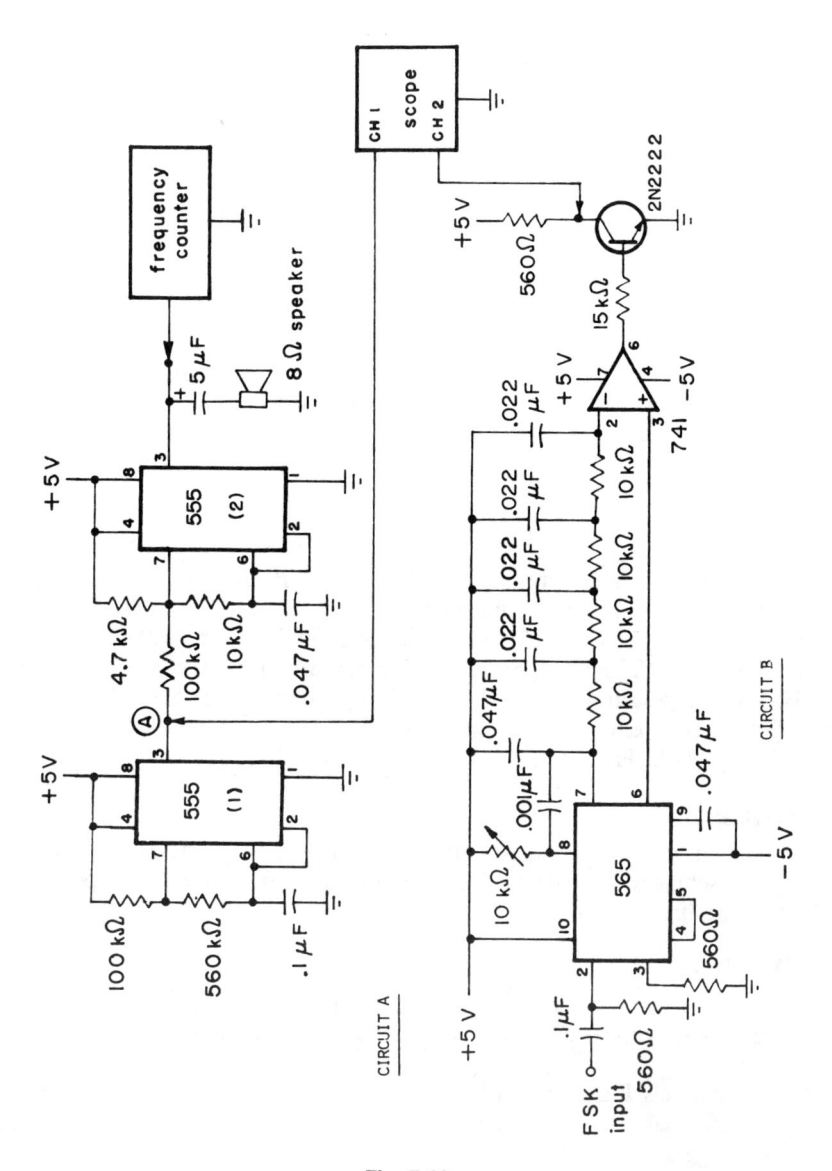

Fig. 7-23.

Step 1

First, wire circuit A (the fsk generator) on one section of the breadboard. Apply power to the breadboard and connect a frequency counter to pin 3 of the No. 2 555 timer. You should hear a sort of "twee-dell" sound that alternates between two different frequencies.

139

Step 2

Next, remove the connection (marked "A") to pin 3 of the No. 1 555 timer. Ground the end of the 100-kΩ resistor that was initially connected to pin 3. Measure the output frequency of the No. 2 timer, which we will call the *mark frequency,* and record your result:

$$f(\text{mark}) = \underline{\qquad} \text{Hz}$$

Step 3

Next, connect the 100-kΩ resistor to the +5-volt supply voltage. You should now hear a steady tone that is *higher* in frequency than before. Measure this output frequency, called the *space frequency,* and record your result:

$$f(\text{space}) = \underline{\qquad} \text{Hz}$$

The frequency difference between the mark and space tones is called the *frequency shift.* As pointed out previously in the discussion of the 565 phase-locked loop, data-communications systems commonly use a 1070-Hz (or 2025-Hz) mark and a 1270 Hz (or 2225-Hz) space, resulting in a 200-Hz shift. Amateur, or "ham" radio teletypewriter systems use frequencies of 2125 Hz and 2295 Hz (170-Hz shift) or 2125 Hz and 2975 Hz (850-Hz shift).

Step 4

Reconnect the 100-kΩ resistor to pin 3 of the No. 1 timer as shown in the schematic diagram. Next, temporarily disconnect the power from the breadboard.

Step 5

Now wire circuit B (the fsk demodulator) as shown in the schematic diagram. Set your oscilloscope for the following settings:

- Channels 1 & 2: 5 V/division
- Time base: 10 ms/division
- Trigger: Channel 1

Step 6

Apply power to the breadboard and connect the output of the fsk generator to the input of the demodulator circuit. Adjust the 10-kΩ potentiometer carefully until the waveforms shown on channels 1 and 2 are the same. At this point, the fsk demodulator is phase-locked to both the mark and space input frequencies. The output of the demodulator circuit is now a logic level that corresponds to the mark and space audio tones.

EXPERIMENT NO. 2

Purpose

The purpose of this experiment is to demonstrate the operation of the 567 phase-locked-loop tone decoder.

Pin Configuration of Integrated-Circuit Chip (Fig. 7-24)

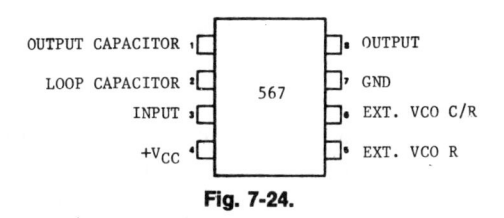

```
OUTPUT CAPACITOR  1 ┤   ├ 8  OUTPUT
LOOP CAPACITOR    2 ┤567├ 7  GND
INPUT             3 ┤   ├ 6  EXT. VCO C/R
+V_CC             4 ┤   ├ 5  EXT. VCO R
```

Fig. 7-24.

Schematic Diagram of Circuit (Fig. 7-25)

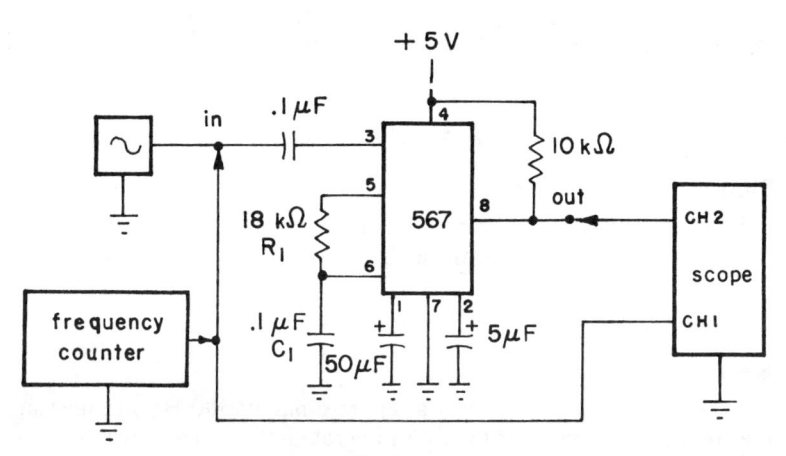

Fig. 7-25.

Step 1

Set your oscilloscope to the following settings:

- Channel 1: 1 V/division
- Channel 2: 5 V/division
- Time base: 0.5 ms/division

Step 2

Wire the circuit shown in the schematic diagram. Apply power to the breadboard. Adjust the frequency generator at 200 Hz with an output voltage of 2 volts peak-to-peak. The output of the tone decoder (channel 2) should be +5 volts (logic 1).

Step 3

Slowly increase the input frequency until the output of the 567 tone decoder changes to logic 0 (0 volts), and record this frequency:

$$f_1 = \underline{\qquad} \text{ Hz}$$

Step 4

Slowly continue to increase the input frequency until the output of the tone decoder returns to +5 volts, and record this frequency:

$$f_2 = \underline{\qquad} \text{ Hz}$$

Step 5

Set the input frequency at about 800 Hz. Slowly decrease the input frequency until the output changes to logic 0, and record this frequency:

$$f_3 = \underline{\qquad} \text{ Hz}$$

Step 6

Slowly continue to decrease the input frequency until the output returns to logic 1, and record this frequency:

$$f_4 = \underline{\qquad} \text{ Hz}$$

Step 7

Now set the input frequency at approximately 500 Hz and measure the frequency at pin 5 of the 567 integrated-circuit chip, which is the vco free-running frequency, f_o. Record your result:

$$f_o = \underline{\qquad} \text{ Hz}$$

From the measurements of Steps 3 through 6, you have determined the range of frequencies for which the 567 tone decoder will lock. On increasing frequencies, lock will occur at f_1 and will stay locked until the input frequency reaches f_2. On decreasing frequencies, lock will occur at f_3 and will stay locked until the input frequency equals f_4. The free-running frequency of the vco is determined by the 18-kΩ resistor (R) and the 0.1-μF capacitor (C) according to the approximate equation:

$$f_o \cong \frac{1.10}{RC}$$

which is about 611 Hz. Within 10%, this should agree with the value you have just determined. The % *bandwidth* is found from

$$\% \text{ bandwidth} = \frac{f_2 - f_4}{f_o} \times 100$$

From your results, compute the % bandwidth and record your result:

$$\% \text{ bandwidth} = \underline{\qquad}$$

For the 567 tone decoder, the % bandwidth is typically 14%. The frequency range, $f_2 - f_4$, is the *lock range* of the decoder phase-locked loop, and is sometimes referred to as the *bandwidth*. The frequency range, $f_3 - f_1$, is the loop *capture range,* and is never greater than the lock range.

Step 8

Starting with an input frequency of 200 Hz, slowly increase the input frequency until you near the frequency that you measured in Step 3 (f_1). Up to this frequency, the vco frequency should remain at the frequency you determined in Step 7. Since the input frequency is outside the loop *lock range,* the loop is not phase-locked and the vco runs at its free-running frequency.

Step 9

Continue to increase the input frequency past f_1. The output of the tone decoder follows the input frequency since the loop is phase-locked. With your frequency counter, compare the input and output frequencies at pins 3 and 5. Are they the same?

Step 10

From the values you determined in Steps 3, 4, 5, and 6, compute the lock range ($f_2 - f_4$) and the capture range ($f_3 - f_1$) for this 567 tone-decoder circuit, and record your results:

$$\text{lock range} = \underline{\qquad} \text{ Hz}$$

$$\text{capture range} = \underline{\qquad} \text{ Hz}$$

Step 11

As an optional exercise, change the resistor between pins 5 and 6 to a different value (4.7 kΩ for example) and repeat the experiment. You should be able to determine the vco center frequency, lock range, and capture range.

EXPERIMENT NO. 3

Purpose

The purpose of this experiment is to demonstrate the operation of the 4046 CMOS phase-locked-loop integrated circuit.

Pin Configuration of Integrated-Circuit Chip (Fig. 7-26)

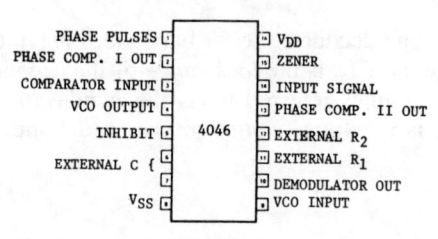

```
        PHASE PULSES  1        16  V_DD
       PHASE COMP. I OUT 2     15  ZENER
       COMPARATOR INPUT  3     14  INPUT SIGNAL
            VCO OUTPUT   4 4046 13  PHASE COMP. II OUT
               INHIBIT   5     12  EXTERNAL R_2
            EXTERNAL C { 6     11  EXTERNAL R_1
                         7     10  DEMODULATOR OUT
                  V_SS   8      9  VCO INPUT
```

Fig. 7-26.

Schematic Diagram of Circuit (Fig. 7-27)

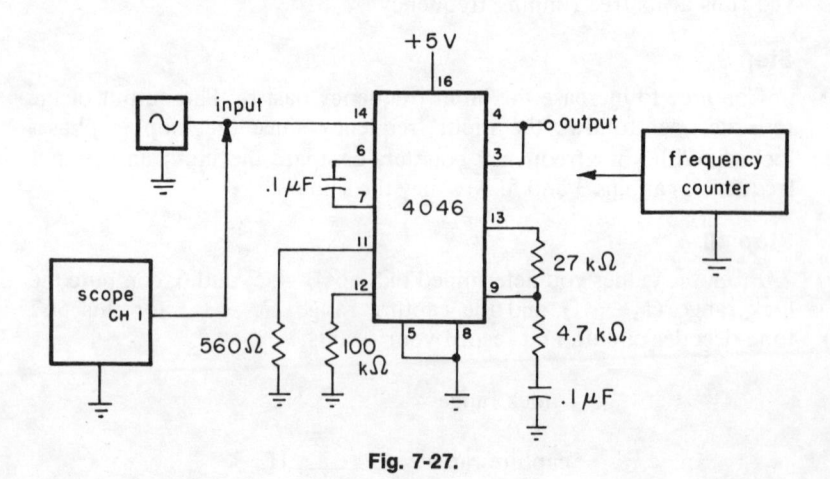

Fig. 7-27.

144

Step 1

Set your oscilloscope for the following settings:

- Channel 1: 0.5 V/division
- Time base: 0.5 ms/division

Step 2

Wire the circuit shown in the schematic diagram and apply power to the breadboard. Adjust the output of the function generator (sine wave) at approximately 1 kHz with the frequency counter, and the peak-to-peak voltage at 1 volt (i.e., 2 vertical divisions). Now connect the frequency counter to pins 3 and 4 of the 4046 device. What do you notice about the output frequency of the phase-locked loop?

The output frequency of the phase-locked loop should be the same as the input.

Step 3

With a piece of wire, connect pin 9 of the 4046 integrated circuit to *ground*. Record the resultant output frequency of the phase-locked loop:

$$f_L = \underline{\qquad} \text{ Hz}$$

This output frequency is the lower range of the vco, which is determined by the 0.1-μF capacitor connected between pins 6 and 7, and the 100-kΩ resistor connected between pin 12 and ground.

Step 4

Now with the same wire, connect pin 9 to the +5-volt supply. You should observe an output frequency that is *higher* than the one you measured in Step 3. Record this frequency:

$$f_H = \underline{\qquad} \text{ Hz}$$

This output frequency is the upper range of the vco, which is determined by the 0.1-μF capacitor connected between pins 6 and 7, and the 560-Ω resistor connected between pin 11 and ground.

Step 5

Now remove the connection between pin 9 and the +5-volt supply. You should again measure an output frequency that is the same as the frequency of the function generator (approximately 1 kHz).

Step 6

Connect the frequency counter to pins 3 and 4 of the 4046 integrated circuit. Now slowly increase the frequency of the function generator. What do you observe on the frequency counter?

You should observe that the output frequency also increases! In fact, the output frequency follows the changes of the input frequency and should be exactly equal. Check the input frequency to confirm this.

Step 7

While watching the output frequency of the phase-locked loop, continue to slowly increase the input frequency and stop when the output frequency does not continue to increase. Measure the input frequency and record your result:

$$f_{in}(H) = \underline{\qquad} Hz$$

You should find that this frequency is about the same as the frequency which you measured in Step 4, *the upper range of the vco*. The phase-locked loop then follows input frequency changes for frequencies below this upper range.

Step 8

Now decrease the input frequency while observing the frequency counter. At some point the output frequency will remain constant. Measure the input frequency and record your result:

$$f_{in}(L) = \underline{\qquad} Hz$$

You should find that this frequency is about the same as the frequency which you measured in Step 3, *the lower range of the vco*. Consequently, the phase-locked-loop circuit follows changes in the input frequency for any frequency between the lower and upper range of the vco. Therefore, the loop is *locked*. The range over which the phase-locked loop follows changes in the input frequency is called the *lock range*. To determine the lock range, subtract the value you determined in Step 8 from the value in Step 7 and record your result:

$$\text{lock range} = \underline{\qquad} Hz$$

The lock range can be changed by simply changing the value of the resistor connected to pin 11 or 12. Decreasing the 100-kΩ resistor at pin 12, for example, *increases* the lower range frequency.

Keep this circuit on your breadboard, as it will be used in the next experiment.

EXPERIMENT NO. 4

Purpose

The purpose of this experiment is to demonstrate a "loss-of-lock" indicator with the phase-locked-loop circuit of Experiment No. 3.

Pin Configuration of Integrated-Circuit Chip (Fig. 7-28)

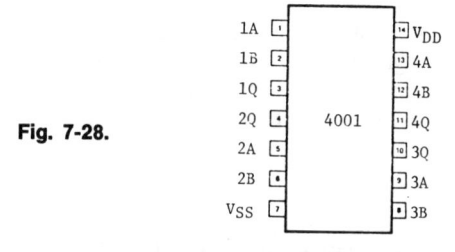

Fig. 7-28.

Schematic Diagram of Circuit (Fig. 7-29)

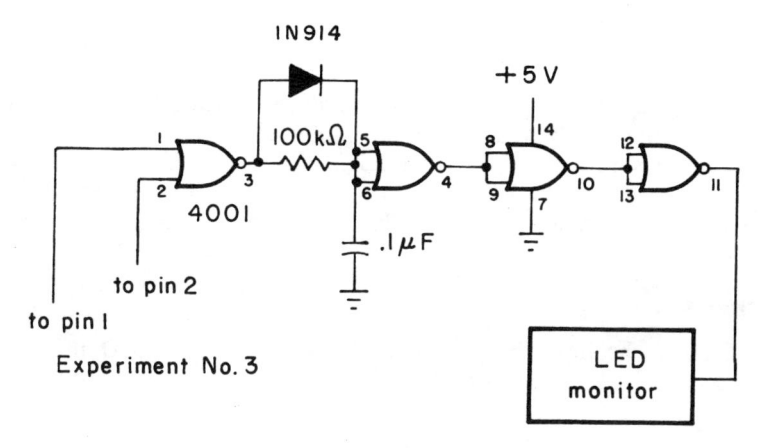

Fig. 7-29.

Step 1

Wire the loss-of-lock circuit shown in the schematic diagram. Connect pin 1 of the 4001 CMOS NOR gate to pin 1 of the 4046 phase-

147

locked loop (PHASE PULSES output of comparator II), and pin 2 of the 4001 NOR gate to pin 2 of the 4046 phase-locked loop (phase comparator I output). Make sure that you have connected the 1N914 diode correctly across the 100-kΩ resistor. The *anode* goes to pin 3 while the *cathode* goes to the junction of pins 5 and 6 of the 4001 NOR gate. The cathode end is usually marked with a colored band.

Step 2

Apply power to the breadboard and set the input at approximately 500 Hz. Is the LED monitor lit or unlit?

The LED monitor should be lit since the 500-Hz input frequency is within the lock range of the loop, which you determined in Steps 7 and 8 of the previous experiment. When the loop is phase-locked, the output of the loss-of-lock circuit (pin 11 of the 4001 NOR gate) is at logic 1.

Step 3

Increase the input frequency just past the upper range of the vco (Step 7 of the previous experiment). What happens to the LED monitor?

The LED monitor should now be *unlit,* indicating that the phase-locked loop is unlocked. The loop is now unlocked since the input frequency is now outside the lock range of the loop. In some cases we have noticed that the LED monitor flickers on and off several times as the loop becomes unlocked. This is due to the transient behavior of the loop.

Step 4

Change the input frequency to 1 kHz. From the LED monitor, is the loop locked or unlocked?

The LED monitor should be lit since the 1-kHz input signal is within the lock range of the loop. We then can use such a circuit with the 4046 CMOS phase-locked loop to visually indicate whether the loop is locked or not.

EXPERIMENT NO. 5

Purpose

The purpose of this experiment is to demonstrate the operation of a 4046 phase-locked loop and a 4017 decade counter as a ×10 frequency multiplier or prescaler.

Pin Configuration of Integrated-Circuit Chips (Fig. 7-30)

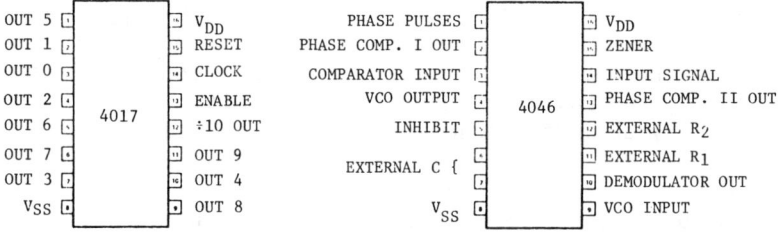

```
     4017                                4046
OUT 5  — V_DD              PHASE PULSES  — V_DD
OUT 1  — RESET           PHASE COMP. I OUT — ZENER
OUT 0  — CLOCK          COMPARATOR INPUT  — INPUT SIGNAL
OUT 2  — ENABLE              VCO OUTPUT  — PHASE COMP. II OUT
OUT 6  — ÷10 OUT              INHIBIT    — EXTERNAL R_2
OUT 7  — OUT 9                            — EXTERNAL R_1
OUT 3  — OUT 4            EXTERNAL C {   — DEMODULATOR OUT
V_SS   — OUT 8                   V_SS    — VCO INPUT
```

Fig. 7-30.

Schematic Diagram of Circuit (Fig. 7-31)

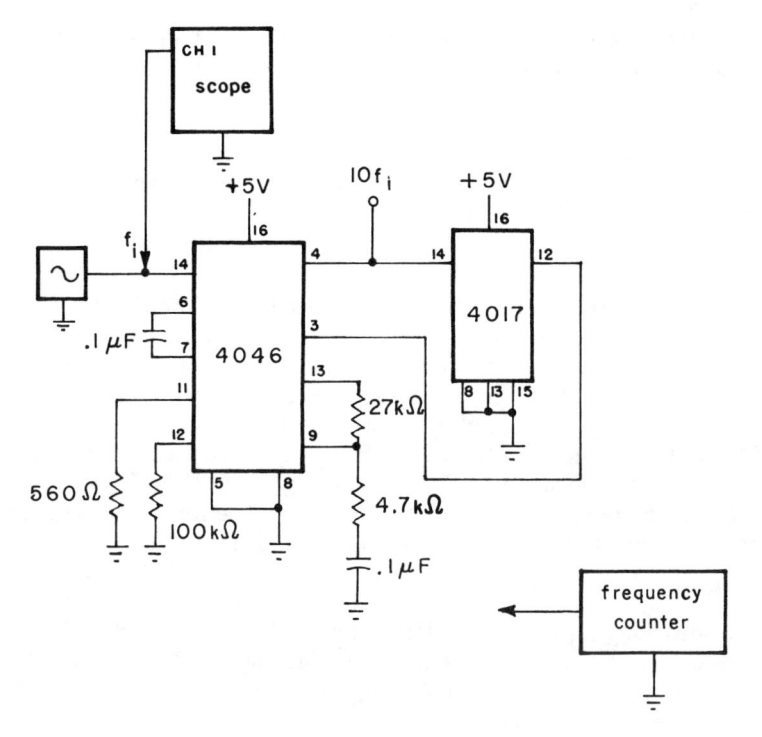

Fig. 7-31.

149

Step 1

Set your oscilloscope for the following settings:

- Channel 1: 0.5 V/division
- Time base: 10 ms/division
- Ac coupling

Step 2

Wire the circuit shown in the schematic diagram and apply power to the breadboard. Connect the frequency counter input to pin 14 of the 4046 integrated circuit. Adjust the function generator so that the input frequency (f_i) is somewhere between 80 and 90 Hz. In addition, adjust the peak-to-peak input voltage at 1 volt.

Step 3

Measure the input frequency and record your result:

$$f_i = _____ \text{ Hz}$$

Step 4

Now connect the frequency counter to pin 4 of the 4046 device. Measure the output frequency and record your result:

$$f_o = _____ \text{ Hz}$$

What relationship do you notice between the frequency that you measured in this step and the one in Step 3?

The output frequency should be 10 times larger than the input. The input frequency that you measured in Step 3 normally has a resolution of ±1 Hz. By using this circuit to multiply the input frequency by 10, we are then able to measure the input frequency with a resolution of ±0.1 Hz. As an example, if you measured an input frequency of 87 Hz, this means that the input frequency could range from 86 to 88 Hz. If the measured output frequency was 867 Hz, the input would be more precisely 86.7 Hz, not 87 Hz! The frequency counter's resolution would then be increased by 1 significant digit.

Step 5

Choose any input frequency between 20 Hz and 300 Hz. Measure both the input and output frequencies of the synthesizer circuit. Over this input frequency range you should be convinced that it is possible to measure an input frequency with a resolution of ±0.1 Hz instead of ±1 Hz.

APPENDIX A

Derivations

THE BASIC TRANSFER SYSTEM

For the basic phase-locked-loop system shown in Fig. A-1, we have a phase detector, a low-pass filter, and a voltage-controlled oscillator, or vco. For a phase difference ($\Delta\phi$) between the input signal and the output of the vco, the output voltage of the phase detector is proportional to this phase difference so that

$$V_o = K_\phi \Delta\phi \qquad \text{(Eq. A-1)}$$

where the constant, K_ϕ, is the conversion gain of the phase detector in V/rad.

In turn, the output voltage of the phase detector is filtered by the low-pass filter, which also determines the dynamic characteristics of the loop. For the time being, the transfer function of the low-pass filter is represented by $F(s)$, since we will consider its specific form shortly. In general, the output of the filter is

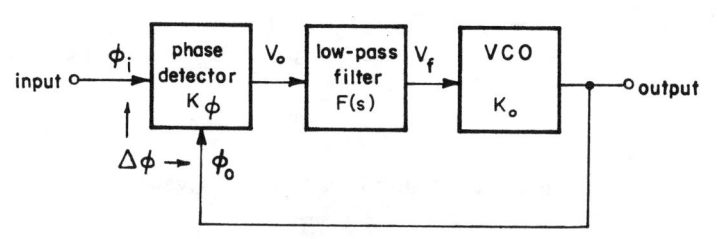

$$V_f(s) = V_o F(s) \qquad \text{(Eq. A-2)}$$

Fig. A-1. Block diagram of the basic phase-locked loop.

The output voltage of the filter then controls the output frequency of the vco. Depending on this voltage, the vco frequency will have a deviation ($\Delta\omega$) from its center frequency (ω_o) so that

$$\Delta\omega(s) = K_o V_f(s) \qquad \text{(Eq. A-3)}$$

where K_o is the conversion gain of the vco in rad/s/V. Since frequency is the time derivative of phase,

$$\omega = \frac{d\phi}{dt} \qquad \text{(Eq. A-4)}$$

Equation A-3 can now be written as

$$\frac{d\phi}{dt} = K_o V_f(s) \qquad \text{(Eq. A-5)}$$

Taking the Laplace transform of Equation A-5,

$$\phi_o(s) = \frac{K_o V_f(s)}{s} \qquad \text{(Eq. A-6)}$$

so that the output signal of the vco is proportional to the integral of the vco input voltage. Using Equations A-1, A-2, and A-6, we can solve for the ratio, $\phi_o(s)/\phi_i(s)$, so that

$$T(s) = \frac{\phi_o(s)}{\phi_i(s)} = \frac{K_\phi K_o F(s)}{s + K_\phi K_o F(s)} \qquad \text{(Eq. A-7)}$$

whose final form, of course, depends on the type of loop filter used.

Loop Filter A

For the simple, passive, low-pass filter shown in Fig. A-2, the transfer function of the network can be written as

$$F_A(s) = \frac{1}{1 + Ts} \qquad \text{(Eq. A-8)}$$

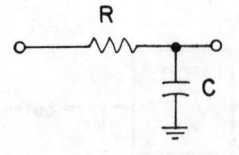

where,
$T = RC$.

Fig. A-2. Passive low-pass filter.

Substitution of Equation A-8 into Equation A-7 gives

$$T_A(s) = \frac{K_\phi K_o/T}{s^2 + (1/T)s + (K_\phi K_o/T)} \qquad \text{(Eq. A-9)}$$

Equating the terms of the denominator of Equation A-9 with the basic characteristic equation of a second-order system,

$$s^2 + 2\zeta\omega_n + \omega_n^2 \qquad \text{(Eq. A-10)}$$

where,
 $\zeta =$ damping factor,
 $\omega_n =$ loop natural frequency,

we find that

$$\zeta = \frac{1}{2}\left(\frac{1}{K_\phi K_o T}\right)^{1/2} \qquad \text{(Eq. A-11)}$$

and

$$\omega_n = \left(\frac{K_\phi K_o}{T}\right)^{1/2} \qquad \text{(Eq. A-12)}$$

so that Equation A-9 can be written in a more convenient form,

$$T_A(s) = \frac{\omega_n^2}{s^2 + 2\zeta\omega_n s + \omega_n^2} \qquad \text{(Eq. A-13)}$$

Loop Filter B

For the passive phase-lag–type filter shown in Fig. A-3, the following transfer function can be written:

$$F_B(s) = \frac{T_2 s + 1}{(T_1 + T_2)s + 1} \qquad \text{(Eq. A-14)}$$

where,
 $T_1 = R_1 C$,
 $T_2 = R_2 C$.

Substitution of Equation A-14 into Equation A-7 yields

$$T_B(s) = \frac{K_\phi K_o[(T_2 s + 1)/(T_1 + T_2)]}{s^2 + [(1 + K_\phi K_o T_2)/(T_1 + T_2)]s + K_\phi K_o/(T_1 + T_2)}$$
$$\text{(Eq. A-15)}$$

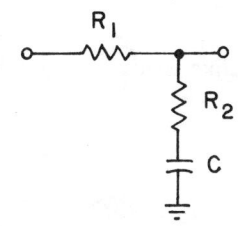

Fig. A-3. Passive phase-lag filter.

153

Equating like terms of Equation A-15 with Equation A-10, we obtain

$$\omega_n = \left(\frac{K_\phi K_o}{T_1 + T_2}\right)^{1/2} \quad \text{(Eq. A-16)}$$

$$\zeta = \frac{1}{2}\left(\frac{K_\phi K_o}{T_1 + T_2}\right)^{1/2}\left[T_2 + \left(\frac{1}{K_\phi K_o}\right)\right] \quad \text{(Eq. A-17)}$$

so that Equation A-15 can be rewritten as

$$T_B(s) = \frac{\omega_n(2\zeta - \omega_n/K_\phi K_o)s + \omega_n^2}{s^2 + 2\zeta\omega_n s + \omega_n^2} \quad \text{(Eq. A-18)}$$

Loop Filter C

For the active filter version of loop filter B, shown in Fig. A-4, the transfer function can be written as

$$F_C(s) = \frac{T_2 s + 1}{sT_1} \quad \text{(Eq. A-19)}$$

where,
$T_1 = R_1 C$,
$T_2 = R_2 C$,

assuming that the amplifier gain is very large. Substitution of Equation A-19 into Equation A-7 gives

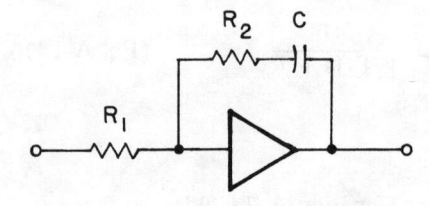

Fig. A-4. Active phase-lag filter.

$$T_C(s) = \frac{\dfrac{K_\phi K_o(1 + sT_2)}{T_1}}{s^2 + \left(\dfrac{K_\phi K_o T_2}{T_1}\right)s + \left(\dfrac{K_\phi K_o}{T_1}\right)} \quad \text{(Eq. A-20)}$$

Equating like terms of Equation A-20 with Equation A-10, we obtain

$$\omega_n = \left(\frac{K_\phi K_o}{T_1}\right)^{1/2} \quad \text{(Eq. A-21)}$$

$$\zeta = \frac{T_2}{2}\left(\frac{K_\phi K_o}{T_1}\right)^{1/2} \quad \text{(Eq. A-22)}$$

so that Equation A-20 can be rewritten as

$$T_C(s) = \frac{2\zeta\omega_n s + \omega_n^2}{s^2 + 2\zeta\omega_n s + \omega_n^2} \quad \text{(Eq. A-23)}$$

which is equal to Equation A-18 if $\omega_n/K_\phi K_o \ll 2\zeta$.

By setting ζ equal to 0 (no damping), and taking the inverse Laplace transform of Equations A-13, A-18, and A-23, we find that

$$T_A(t) = T_C(t) = \left(\frac{B^2 + \omega_n^2}{\omega_n}\right)^{1/2} \sin(\omega_n t + \theta)$$

(Eq. A-24)

where,
$B = K_\phi K_o$,
$\theta = \tan^{-1}(B/\omega_n)$,

and,

$$T_B(t) = \omega_n \sin(\omega_n t). \quad \text{(Eq. A-25)}$$

For all three loop filters, the phase-locked-loop system degenerates into a sinusoidal oscillator having a natural frequency of ω_n.

DERIVATION OF LOOP BANDWIDTH

Assuming the type-B loop filter (Equation A-18), the substitution of $s = j\omega$ into Equation A-18 yields

$$T_B(j\omega) = \frac{\omega_n^2 + j2\zeta\omega_n\omega}{(\omega_n^2 - \omega^2) + j2\zeta\omega_n\omega} \quad \text{(Eq. A-26)}$$

To determine the 3-dB bandwidth ($\omega = \omega_{3dB}$), we set

$$\left|T_B(j\omega)\right|^2 = \frac{1}{2} \quad \text{(Eq. A-27)}$$

so that

$$\omega^4 - \omega^2[2\omega_n^2(2\zeta^2 + 1)] - \omega_n^4 = 0 \quad \text{(Eq. A-28)}$$

Since $\omega = \omega_{3dB}$, Equation A-28 can be factored, giving

$$\frac{\omega_{3dB}}{\omega_n} = \{2\zeta^2 + 1 + [(2\zeta^2 + 1)^2 + 1]^{1/2}\}^{1/2} \quad \text{(Eq. A-29)}$$

GRAPHICAL DETERMINATION OF DAMPING FACTOR

The function with time of the damped sinusoidal waveform shown in Fig. A-5 can be expressed as

$$y(t) = \left(\frac{y_o}{\omega_d}\right) e^{-\zeta \omega_n t} \sin(\omega_d t) \qquad \text{(Eq. A-30)}$$

where,
y_o = y intercept at t = 0,
ω_n = undamped natural frequency,
ω_d = damped natural frequency,
  = $\omega_n (1 - \zeta^2)^{1/2}$.

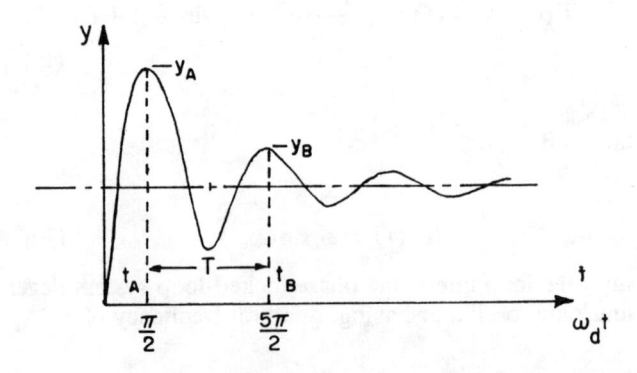

Fig. A-5. Damped sinusoidal waveform.

Equation A-30 can be written in terms of the damped natural frequency, so that

$$y(t) = \left(\frac{y_o}{\omega_d}\right) e^{-[\zeta/(1-\zeta^2)^{1/2}]\omega_d t} \sin(\omega_d t) \qquad \text{(Eq. A-31)}$$

At time $t = t_A$, $\omega_d t = \pi/2$ radians (90°), so that

$$y_A = \left(\frac{y_o}{\omega_d}\right) e^{-[\zeta/(1-\zeta^2)^{1/2}](\pi/2)} \sin(\pi/2) \qquad \text{(Eq. A-32)}$$

Likewise, at time $t = t_B$, $\omega_d t = 5\pi/2$ radians (450°), so that

$$y_B = \left(\frac{y_o}{\omega_d}\right) e^{-[\zeta/(1-\zeta^2)^{1/2}](5\pi/2)} \sin(5\pi/2) \qquad \text{(Eq. A-33)}$$

Dividing Equation A-32 by A-33,

$$\frac{y_A}{y_B} = e^{[\zeta/(1-\zeta^2)^{1/2}](2\pi)} \qquad \text{(Eq. A-34)}$$

since $\sin(\pi/2) = \sin(5\pi/2)$. Taking the natural logarithm of Equation A-34, we obtain

$$\ln(y_A/y_B) = (2\pi)\left(\frac{\zeta}{(1-\zeta^2)^{1/2}}\right) \qquad \text{(Eq. A-35)}$$

Solving for ζ, we find that

$$\zeta = \frac{\gamma}{(1+\gamma^2)^{1/2}} \quad \text{(Eq. A-36)}$$

where,
$\gamma = (1/2\pi)\ln(y_A/y_B)$

Therefore, the damping factor can be determined solely by knowing the peak amplitude of *two consecutive positive peaks which are exactly 1 cycle apart.*

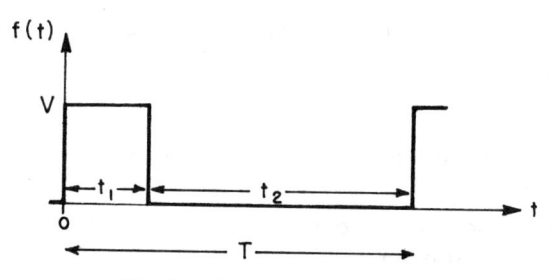

Fig. A-6. Periodic pulse train.

AVERAGE VALUE OF A PULSE TRAIN

The average (dc) value of any periodic waveform is given by

$$\text{average value} = \frac{1}{T}\int_0^T f(t)\,dt \quad \text{(Eq. A-37)}$$

For the periodic pulse train shown in Fig. A-6, we have

$$\text{average value} = \frac{1}{T}\left(\int_0^{t_1} V\,dt + \int_{t_1}^{t_2} 0\,dt\right) \quad \text{(Eq. A-38)}$$

$$= \left(\frac{t_1}{T}\right)V \quad \text{(Eq. A-39)}$$

However, the ratio t_1/T is the duty cycle (D) of the pulse train, so that Equation A-39 can be restated as

$$\text{average value} = VD \quad \text{(Eq. A-40)}$$

so that the average value of a periodic pulse train is directly proportional to the waveform's duty cycle.

APPENDIX B

Data Sheets

This appendix contains the data sheets of the following devices:
1. 560 Phase-Locked Loop
2. 561 Phase-Locked Loop
3. 562 Phase-Locked Loop
4. 564 Phase-Locked Loop
5. 565 Phase-Locked Loop
6. 567 Tone Decoder Phase-Locked Loop
7. 4046 COS/MOS Phase-Locked Loop
8. MC1648 Voltage-Controlled Oscillator
9. MC4024 Voltage-Controlled Multivibrator
10. MC4044 Phase-Frequency Detector
11. HCTR 0320 CMOS Digital Frequency Synthesizer

These data sheets are reproduced with permission of the following manufacturers:

1. Hughes Aircraft Company
   Solid State Products Division
   500 Superior Avenue
   Newport Beach, CA 92663

2. Motorola Semiconductor Products, Inc.
   P.O. Box 20912
   Phoeniz, AZ 85036

3. RCA Solid State Division
   Route 202
   Somerville, NJ 08876

4. Signetics Corporation
   811 East Arques Avenue
   Sunnyvale, CA 94086

PHASE LOCKED LOOP 560

LINEAR INTEGRATED CIRCUITS

DESCRIPTION

The NE560B Phase Locked Loop (PLL) is a monolithic signal conditioner, and demodulator system comprising a VCO, Phase Comparator, Amplifier and Low Pass Filter, interconnected as shown in the accompanying block diagram. The center frequency of the PLL is determined by the free running frequency (f_o) of the VCO. This VCO frequency is set by an external capacitor and can be fine tuned by an optional Potentiometer. The low pass filter, which determines the capture characteristics of the loop, is formed by the two capacitors and two resistors at the Phase Comparator output.

The PLL system has a set of self biased inputs which can be utilized in either a differential or single ended mode. The VCO output, in differential form, is available for signal conditioning frequency synchronization, multiplication and division applications. Terminals are provided for optional extended control of the tracking range, VCO frequency, and output DC level.

The monolithic signal conditioner-demodulator system is useful over a wide range of frequencies from less than 1 Hz to more than 15 MHz with an adjustable tracking range of ±1% to ±15%.

FEATURES
- FM DEMODULATION WITHOUT TUNED CIRCUITS
- NARROW BANDPASS - TO ± 1% ADJUSTABLE
- TRACKING RANGE
- EXACT FREQUENCY DUPLICATION IN HIGH
- NOISE ENVIRONMENT
- WIDE TRACKING RANGE ±15%
- HIGH LINEARITY - 1% DISTORTION MAX
- FREQUENCY MULTIPLICATION AND DIVISION
- THROUGH HARMONIC LOCKING

APPLICATIONS
TONE DECODERS
FM IF STRIPS
TELEMETRY DECODERS
DATA SYNCHRONIZERS
SIGNAL RECONSTITUTION
SIGNAL GENERATORS
MODEMS
TRACKING FILTERS
SCA RECEIVERS
FSK RECEIVERS
WIDE BAND HIGH LINEARITY DETECTORS

ABSOLUTE MAXIMUM RATINGS

Maximum Operating Voltage	26V
Input Voltage	1V Rms
Storage Temperature	-65°C to 150°C
Operating Temperature	0°C to 70°C
Power Dissipation	300 mw

Limiting values above which serviceability may be impaired

PIN CONFIGURATION

B PACKAGE
(Top View)

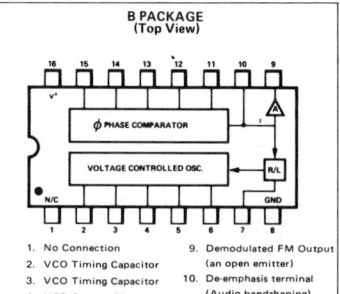

1. No Connection
2. VCO Timing Capacitor
3. VCO Timing Capacitor
4. VCO Output #2
5. VCO Output #1
6. Fine Tuning
7. Range Control
8. Ground (or Negative Power Supply)
9. Demodulated FM Output (an open emitter)
10. De-emphasis terminal (Audio bandshaping)
11. Offset Adjustment
12. FM/RF Input #1
13. FM/RF Input #2
14. Low Pass Loop Filter
15. Low Pass Loop Filter
16. Positive Power Supply

ORDER PART NO. NE560B

BLOCK DIAGRAM

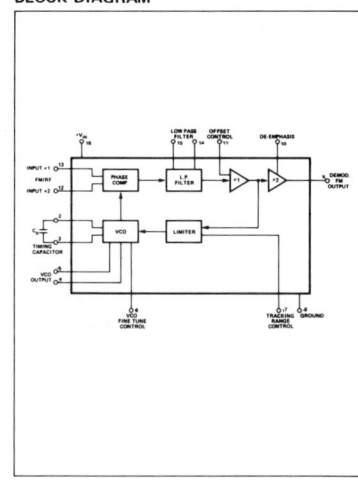

Courtesy Signetics Corp.

560 — PHASE LOCKED LOOP

GENERAL ELECTRICAL CHARACTERISTICS
(15KΩ Pin 9 to GND, Input Pin 12 or Pin 13 AC Ground Unused Input, Optional Controls Not Connected, V+ = 18V Unless Otherwise Specified $T_A = 25°C$)

CHARACTERISTICS	MIN	TYP	MAX	UNITS	TEST CONDITIONS
Lowest Practical Operating Frequency		0.1		Hz	
Maximum Operating Frequency	15	30		MHz	
Supply Current	7	9	11	Ma	
Minimum Input Signal for Lock		100		µV	
Dynamic Range		60		dB	
VCO Temp Coefficient*		±0.06	±0.12	%/°C	Measured at 2 MHz, with both inputs AC grounded
VCO Supply Voltage Regulation		±0.3	±2	%/V	Measured at 2 MHz
Input Resistance		2		KΩ	
Input Capacitance		4		Pf	
Input DC Level		+4		V	
Output DC Level	+12	+14	+16	V	
Available Output Swing		4		V_{p-p}	Measured at Pin 9
AM Rejection*	30	40		dB	See Figure 1
De-emphasis Resistance		8		KΩ	

*ACC Test Sub Group C.

ELECTRICAL CHARACTERISTICS (For FM Applications, Figure 2)
(15KΩ Pin 9 to GND, Input Pin 12 or 13, AC Ground Unused Input, Optional Controls Not Connected, V+ = 18V Unless Otherwise Specified $T_A = 25°C$)

CHARACTERISTICS	MIN	TYP	MAX	UNITS	TEST CONDITIONS
10.7 MHz Operation Deviation 75 kHz Source Impedance = 50Ω					
Detection Threshold		120	300	µV	
Demodulated Output Amplitude	30	60		mV	V_{in} = 1 mv Rms Modulation Frequency 1 kHz
Distortion*		.3	1	% T.H.D.	V_{in} = 1 mv Rms Modulation Frequency 1 kHz
Signal to Noise Ratio S+N/N		35		dB	V_{in} = 1 mv Rms Modulation Frequency 1 kHz
4.5 MHz Operation Deviation = 25 kHz, Source Impedance = 50Ω					
Detection Threshold		120	300	µV	
Demodulated Output Amplitude	30	60		mV	V_{in} = 1 mv Rms Modulation Frequency 1 kHz
Distortion		0.3	1.0	% T.H.D.	V_{in} = 1 mv Rms Modulation Frequency 1 kHz
Signal to Noise Ratio S+N/N		35		dB	V_{in} = 1 mv Rms Modulation Frequency 1 kHz
Wide Deviation $\Delta F/f_o$ = 5% Input = 4.5 MHz Deviation = 225 kHz @ 1 kHz Modulation Rate					
Detection Threshold		1	5	mV	
Demodulated Output	0.2	0.5		Vrms	V_{in} = 5 mv Rms
Distortion		0.8		% T.H.D.	V_{in} = 5 mv Rms
Signal to Noise Ratio S+N/N		50		dB	V_{in} = 5 mv Rms

*ACC Test Sub Group C.

ELECTRICAL CHARACTERISTICS (For Tracking Filter, Figure 3)
(15KΩ Pin 9 to GND, Input Pin 12 or Pin 13 AC Ground Unused Input, Optional Controls Not Connected, V+ = 18V Unless Otherwise Specified $T_A = 25°C$)

CHARACTERISTICS	MIN	TYP	MAX	UNITS	TEST CONDITIONS
Tracking Range	±5	±15		% of f_o	V_{in} = 5 mv Rms
Minimum Signal to Sustain Lock 0°C to 70°C		0.8		mv Rms	Input 2 MHz - See Characteristic Curves
VCO Output Impedance		1		kΩ	
VCO Output Swing	0.4	0.6		V_{p-p}	Input 2 MHz Measured with high impedance Probe with less than 10 Pf Capacitance
VCO Output DC Level		+6.5		V	
Side Band Suppression		35		dB	Input 2 MHz with ±100 kHz Side Band Separation and 3 kHz Low Pass Filter Input 1 mv Peak for Carrier Each Side Band $C_1 = 0.01 \mu F$ $R_1 = 0$

Courtesy Signetics Corp.

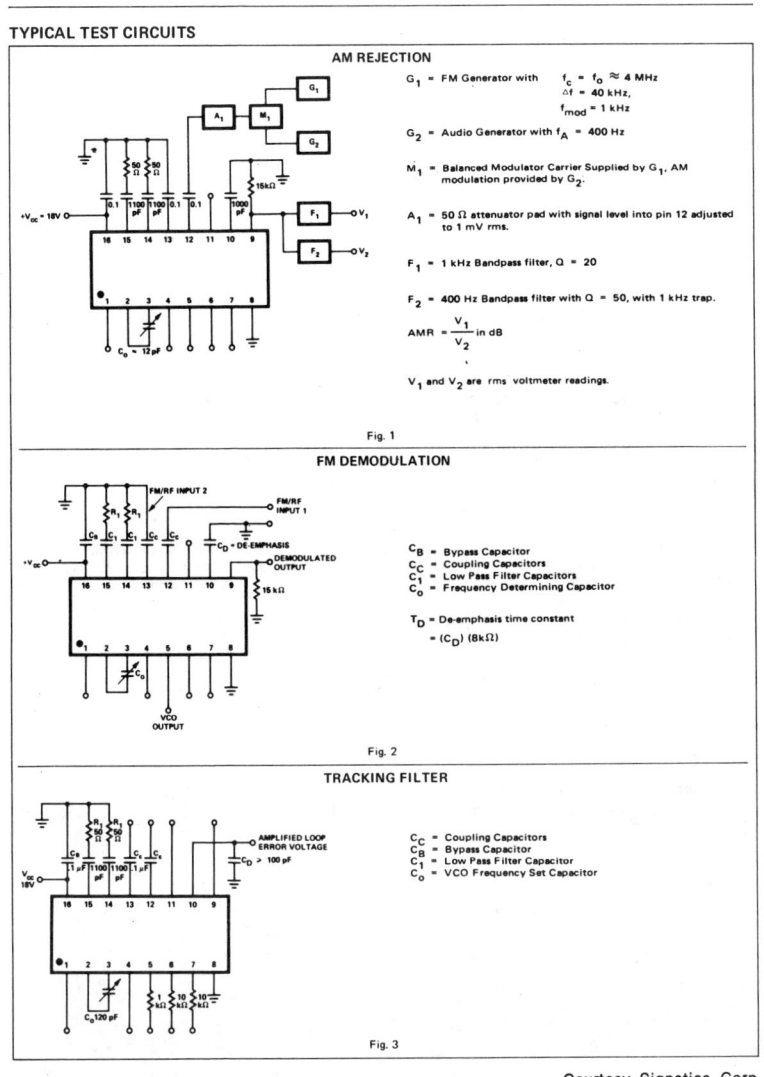

560 – PHASE LOCKED LOOP

TYPICAL TEST CIRCUITS

AM REJECTION

G_1 = FM Generator with $f_c = f_o \approx 4$ MHz, $\Delta f = 40$ kHz, $f_{mod} = 1$ kHz

G_2 = Audio Generator with $f_A = 400$ Hz

M_1 = Balanced Modulator Carrier Supplied by G_1, AM modulation provided by G_2.

A_1 = 50 Ω attenuator pad with signal level into pin 12 adjusted to 1 mV rms.

F_1 = 1 kHz Bandpass filter, Q = 20

F_2 = 400 Hz Bandpass filter with Q = 50, with 1 kHz trap.

$$AMR = \frac{V_1}{V_2} \text{ in dB}$$

V_1 and V_2 are rms voltmeter readings.

Fig. 1

FM DEMODULATION

C_B = Bypass Capacitor
C_C = Coupling Capacitors
C_1 = Low Pass Filter Capacitors
C_0 = Frequency Determining Capacitor

T_D = De-emphasis time constant
= $(C_D)(8k\Omega)$

Fig. 2

TRACKING FILTER

C_C = Coupling Capacitors
C_B = Bypass Capacitor
C_1 = Low Pass Filter Capacitor
C_0 = VCO Frequency Set Capacitor

Fig. 3

Courtesy Signetics Corp.

161

560 — PHASE LOCKED LOOP

TYPICAL CHARACTERISTIC CURVES

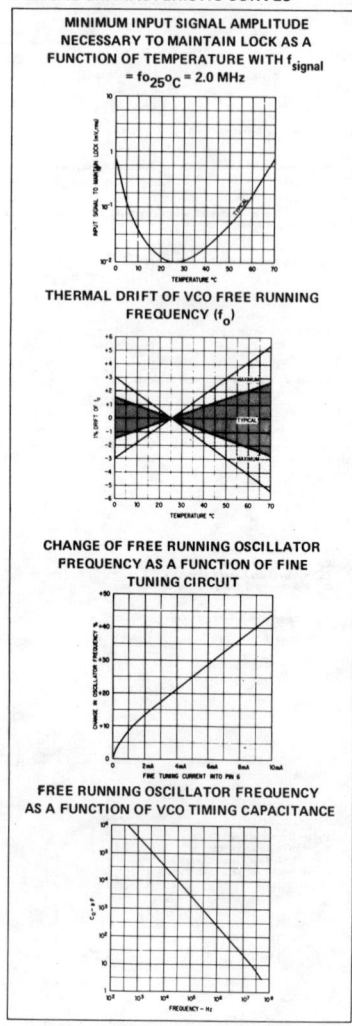

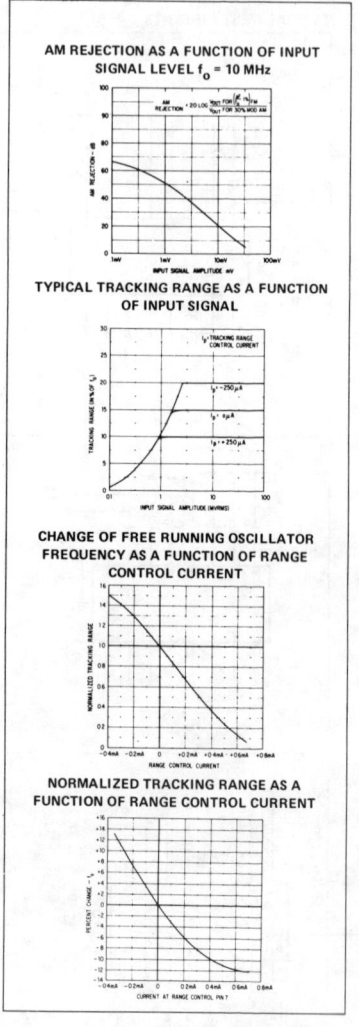

MINIMUM INPUT SIGNAL AMPLITUDE NECESSARY TO MAINTAIN LOCK AS A FUNCTION OF TEMPERATURE WITH $f_{signal} = fo_{25°C} = 2.0$ MHz

AM REJECTION AS A FUNCTION OF INPUT SIGNAL LEVEL $f_o = 10$ MHz

THERMAL DRIFT OF VCO FREE RUNNING FREQUENCY (f_o)

TYPICAL TRACKING RANGE AS A FUNCTION OF INPUT SIGNAL

CHANGE OF FREE RUNNING OSCILLATOR FREQUENCY AS A FUNCTION OF FINE TUNING CIRCUIT

CHANGE OF FREE RUNNING OSCILLATOR FREQUENCY AS A FUNCTION OF RANGE CONTROL CURRENT

FREE RUNNING OSCILLATOR FREQUENCY AS A FUNCTION OF VCO TIMING CAPACITANCE

NORMALIZED TRACKING RANGE AS A FUNCTION OF RANGE CONTROL CURRENT

Courtesy Signetics Corp.

560 — PHASE LOCKED LOOP

EXTERNAL CONTROLS

1. **Loop Low Pass Filter (Pins 14 and 15)**
   The equivalent circuit for the loop low-pass filter can be represented as:

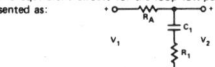

   where RA (6K Ω) is the effective resistance seen looking into Pin #14 or Pin #15.
   The corresponding filter transfer characteristics are:

   $$\frac{V_2}{V_1}(S) = (S) = \frac{1 + S R_1 C_1}{1 + S(R_1 + R_A) C_1}$$

   where S is the complex frequency variable.

2. **Loop Gain (Threshold) Control**
   The overall Phase Locked Loop gain can be reduced by connecting a feedback resistor, R_F, across the low-pass filter terminals, Pins #14 and #15. This causes the loop gain and the detection sensitivity to decrease by a factor α ($\alpha < 1$) where:

   $$\alpha = \frac{R_F}{2 R_A + R_F}$$

   Reduction of loop gain may be desirable at high input signal levels ($V_{in} > 30$ mV) and at high frequencies ($f_o > 5$ MHz) where excessively high loop gain may cause instability.

3. **Tracking Range Control (Pin 7)**
   Any bias current, I_p, injected into the tracking range control, reduces the tracking range of the PLL by decreasing the output of the limiter. The variation of the tracking range and the center frequency, as a function of I_p, are shown in the characteristic curves with I_p defined positive going into the tracking range control terminal. This terminal is normally at a DC level of +0.6 Volts and presents an impedance of 600 Ω.

4. **External Fine Tuning (Pin 6)**
   Any bias current injected into the fine tuning terminal increases the frequency of oscillation, f_o, as shown in the characteristic curves. This current is defined Positive into the fine tuning terminal. This terminal is at a typical DC level of +1.3 Volts and has a dynamic impedance of 100Ω to ground.

5. **Offset Adjustment (Pin 11)**
   Application of a bias voltage to the offset adjustment terminal modifies the current in the output amplifier setting the DC level at the output. The effect on the loop is to modify the relationship between the VCO free running frequency and the lock range, allowing the VCO free running frequency to be positioned at different points throughout the lock range.

   Nominally this terminal is at +4V DC and has an input impedance of 3K Ω. The offset adjustment is optional. The characteristics specified correspond to operation of the circuit with this terminal open circuited.

6. **De-emphasis Filter (Pin 10)**
   The de-emphasis terminal is normally used when the PLL is used to demodulate Frequency Modulated Audio signals. In this application, a capacitor from this terminal to ground provides the required de-emphasis. For other applications, this terminal may be used for band shaping the output signal. The 3 dB bandwidth of the output amplifier in the system block diagram (see Figure 2.) is related to the de-emphasis capacitor, C_D, as:

   $$f_{3dB} = \frac{1}{2 R_a C_D}$$

   where R_D is the 8000 ohm resistance seen looking into the de-emphasis terminal.
   When the PLL system is utilized for signal conditioning, and the loop error voltage is not utilized, de-emphasis terminal should be AC grounded.

Courtesy Signetics Corp.

163

PHASE LOCKED LOOP 561

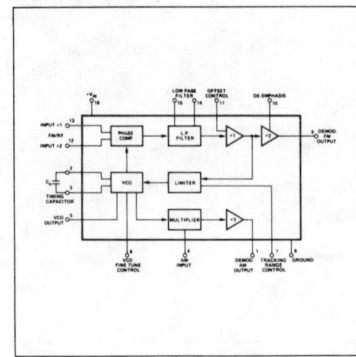

LINEAR INTEGRATED CIRCUITS

DESCRIPTION

The NE561B Phase Locked Loop (PLL) is a monolithic signal conditioner, and demodulator system comprising a VCO, Phase Comparator, Amplifier and Low Pass Filter, interconnected as shown in the accompanying block diagram. The center frequency of the PLL is determined by the free running frequency (f_o) of the VCO. This VCO frequency is set by an external capacitor and can be fine tuned by an optional Potentiometer. The low pass filter, which determines the capture characteristics of the loop is formed by the two capacitors and two resistors at the Phase Comparator output.

The PLL system has a set of self biased inputs which can be utilized in either a differential or single ended mode. The VCO output is available for signal conditioning, frequency synchronization, multiplication and division applications. Terminals are provided for optional external control of the tracking range, VCO frequency, and output DC level. An analog multiplier block is incorporated into the PLL system to provide frequency selective synchronous AM detection capability.

The monolithic signal conditioner-demodulator system is useful over a wide range of frequencies from less than 1 Hz to more than 15 MHz with an adjustable tracking range of $\pm 1\%$ to $\pm 15\%$.

FEATURES
- FM DEMODULATION WITHOUT TUNED CIRCUITS
- SYNCHRONOUS AM DETECTION
- NARROW BAND PASS TO $\pm 1\%$
- EXACT FREQUENCY DUPLICATION IN HIGH NOISE ENVIRONMENT
- ADJUSTABLE TRACKING RANGE
- WIDE TRACKING RANGE $\pm 15\%$
- HIGH LINEARITY - 1% DISTORTION MAX
- FREQUENCY MULTIPLICATION AND DIVISION THROUGH HARMONIC LOCKING

BLOCK DIAGRAM

PIN CONFIGURATION
B PACKAGE

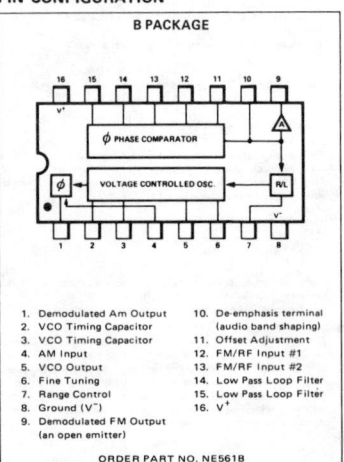

1. Demodulated Am Output
2. VCO Timing Capacitor
3. VCO Timing Capacitor
4. AM Input
5. VCO Output
6. Fine Tuning
7. Range Control
8. Ground (V⁻)
9. Demodulated FM Output (an open emitter)
10. De-emphasis terminal (audio band shaping)
11. Offset Adjustment
12. FM/RF Input #1
13. FM/RF Input #2
14. Low Pass Loop Filter
15. Low Pass Loop Filter
16. V⁺

ORDER PART NO. NE561B

ABSOLUTE MAXIMUM RATINGS
Maximum Operating Voltage	26V
Input Voltage	1V RMS
Storage Temperature	$-65°C$ to $150°C$
Operating Temperature	$0°C$ to $70°C$
Power Dissipation	300mW

Limiting values above which serviceability may be imparied

APPLICATIONS
TONE DECODERS
AM-FM-IF STRIPS
TELEMETRY DECODERS
DATA SYNCHRONIZERS
SIGNAL RECONSTITUTION
SIGNAL GENERATORS
MODEMS
TRACKING FILTERS
SCA RECEIVERS
FSK RECEIVERS
WIDE BAND HIGH LINEARITY DETECTORS
SYNCHRONOUS DETECTORS
AM RECEIVER

Courtesy Signetics Corp.

561 – PHASE LOCKED LOOP

GENERAL ELECTRICAL CHARACTERISTICS
(15KΩ Pin 9 to GND, Input Pin 12 or Pin 13 AC Ground Unused Input, Optional Controls Not Connected, V+ = 18V Unless Otherwise Specified $T_A = 25°C$)

CHARACTERISTICS	MIN	TYP	MAX	UNITS	TEST CONDITIONS
Lowest Practical Operating Frequency		0.1		Hz	
Maximum Operating Frequency	15	30		MHz	
Supply Current	8	10	12	Ma	
Minimum Input Signal for Lock		100		µV	
Dynamic Range		60		dB	
VCO Temp Coefficient*		±0.06	±0.12	%/°C	Measured at 2 MHz, with both inputs AC grounded
VCO Supply Voltage Regulation		±0.3	±2	%/V	Measured at 2 MHz
Input Resistance		2		kΩ	
Input Capacitance		4		pF	
Input DC Level		+4		V	
Output DC Level	+12	+14	+16	V	
Available Output Swing		4		V_{p-p}	Measured at Pin 9
AM Rejection*	30	40		dB	See Figure 3
De-emphasis Resistance		8		kΩ	

*ACC Test Sub Group C.

ELECTRICAL CHARACTERISTICS (For FM Applications, Figure 2) (15KΩ Pin 9 to GND, Input Pin 12 or 13, AC Ground Unused Input, Optional Controls Not Connected, V+ = 18V Unless Otherwise Specified $T_A = 25°C$)

CHARACTERISTICS	MIN	TYP	MAX	UNITS	TEST CONDITIONS
10.7 MHz Operation Deviation 75 kHz Source Impedance = 50Ω					
Detection Threshold		120	300	µV	
Demodulated Output Amplitude	30	60		mV	Vin = 1 mv Rms Modulation Frequency 1 kHz
Distortion*		.3	1	% T.H.D.	Vin = 1 mv Rms Modulation Frequency 1 kHz
Signal to Noise Ratio $\frac{S+N}{N}$		35		dB	Vin = 1 mv Rms Modulation Frequency 1 kHz
4.5 MHz Operation Deviation = 25 kHz, Source Impedance = 50Ω					
Detection Threshold		120	300	µV	
Demodulated Output Amplitude	30	60		mV	Vin = 1 mv Rms Modulation Frequency 1 kHz
Distortion		0.3	1.0	% T.H.D.	Vin = 1 mv Rms Modulation Frequency 1 kHz
Signal to Noise Ratio $\frac{S+N}{N}$		35		dB	Vin = 1 mv Rms Modulation Frequency 1 kHz
Wide Deviation $\Delta F/f_o = 5\%$ Input = 4.5 MHz Deviation = 225 kHz @ 1 kHz modulation rate					
Detection Threshold		1	5	mV	
Demodulated Output	0.2	0.5		Vrms	Vin = 5 mv Rms
Distortion		0.8		% T.H.D.	Vin = 5 mv Rms
Signal to Noise Ratio $\frac{S+N}{N}$		50		dB	Vin = 5 mv Rms

*ACC Test Sub Group C.

ELECTRICAL CHARACTERISTICS (For Tracking Filter, Figure 1) (15KΩ Pin 9 to GND, Input Pin 12 or Pin 13 AC Ground Unused Input, Optional Controls Not Connected, V+ = 18V Unless Otherwise Specified $T_A = 25°C$)

CHARACTERISTICS	MIN	TYP	MAX	UNITS	TEST CONDITIONS
Tracking Range	±5	±20		% of f_o	Vin 5 mv Rms
Minimum Signal to Sustain Lock 0°C to 70°C		0.8		mv Rms	Input 2 MHz - See Characteristic Curves
VCO Output Impedance		1		kΩ	
VCO Output Swing	0.4	0.6		V_{p-p}	Input 2 MHz Measured with high impedance. Probe with less than 10 pF capacitance.
VCO Output DC Level		+6.5		V	
Side Band Suppression		35		dB	Input 2 MHz with ±100 kHz Sideband Separation and 3 kHz Low Pass Filter. Input 1 mv Peak for Carrier and each Sideband $C_1 = 0.01 \mu F$ $R_1 = 0$

Courtesy Signetics Corp.

561 – PHASE LOCKED LOOP

ELECTRICAL CHARACTERISTICS (For AM Synchronous Detector, Figure 4) (15KΩ Pin 9 to GND, Input Pin 12 or Pin 13 AC Ground Unused Input, Optional Controls Not Connected, V+ = 18V Unless Otherwise Specified $T_A = 25°C$)

CHARACTERISTICS	MIN	TYP	MAX	UNITS	TEST CONDITIONS
Input Impedance		3		kΩ	
Output Impedance		8		kΩ	
Output DC Level	+10	+14	+17	V	
AM Conversion Gain	3	12		dB	See Definition of Terms
Out of Band Rejection		30		dB	See Definition of Terms
Distortion		1		% T.H.D.	

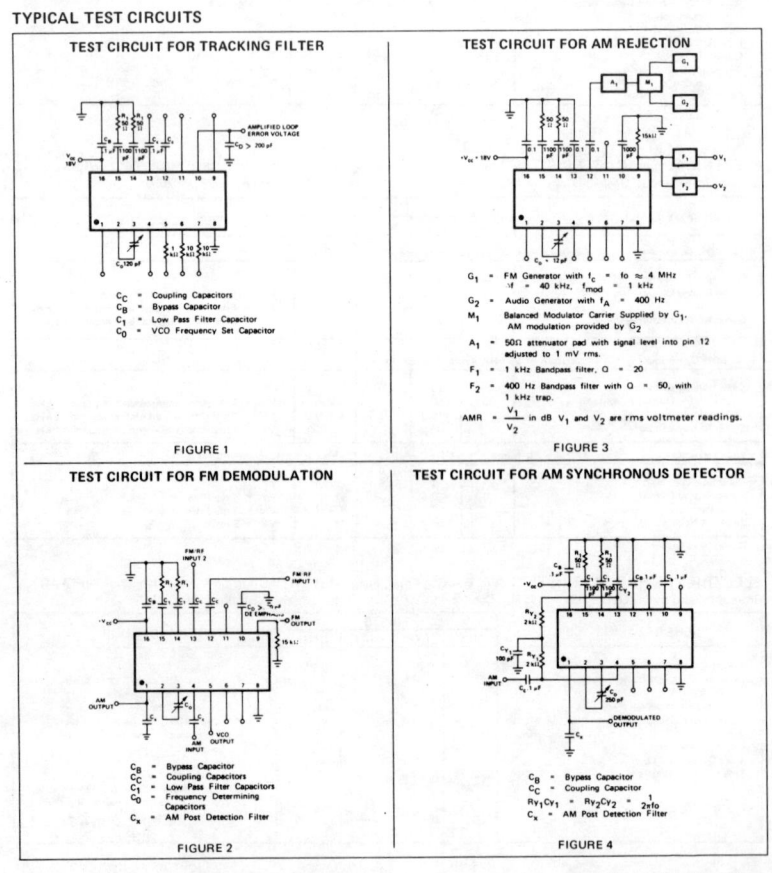

TYPICAL TEST CIRCUITS

TEST CIRCUIT FOR TRACKING FILTER

C_C = Coupling Capacitors
C_B = Bypass Capacitor
C_1 = Low Pass Filter Capacitor
C_0 = VCO Frequency Set Capacitor

FIGURE 1

TEST CIRCUIT FOR AM REJECTION

G_1 = FM Generator with f_c = f_0 ≈ 4 MHz, Δf = 40 kHz, f_{mod} = 1 kHz
G_2 = Audio Generator with f_A = 400 Hz
M_1 = Balanced Modulator Carrier Supplied by G_1, AM modulation provided by G_2
A_1 = 50Ω attenuator pad with signal level into pin 12 adjusted to 1 mV rms.
F_1 = 1 kHz Bandpass filter, Q = 20
F_2 = 400 Hz Bandpass filter with Q = 50, with 1 kHz trap.
AMR = $\frac{V_1}{V_2}$ in dB V_1 and V_2 are rms voltmeter readings.

FIGURE 3

TEST CIRCUIT FOR FM DEMODULATION

C_B = Bypass Capacitor
C_C = Coupling Capacitors
C_1 = Low Pass Filter Capacitors
C_0 = Frequency Determining Capacitors
C_x = AM Post Detection Filter

FIGURE 2

TEST CIRCUIT FOR AM SYNCHRONOUS DETECTOR

C_B = Bypass Capacitor
C_C = Coupling Capacitor
$R_{V_1}C_{V_1}$ = $R_{V_2}C_{V_2}$ = $\frac{1}{2\pi f_0}$
C_x = AM Post Detection Filter

FIGURE 4

Courtesy Signetics Corp.

561 — PHASE LOCKED LOOP

TYPICAL CHARACTERISTIC CURVES

MINIMUM INPUT SIGNAL AMPLITUDE NECESSARY TO MAINTAIN LOCK AS A FUNCTION OF TEMPERATURE WITH $f_{signal} = fo_{25°C} = 2.0$ MHz

AM REJECTION AS A FUNCTION OF INPUT SIGNAL LEVEL $f_o = 10$ MHz

THERMAL DRIFT OF VCO FREE RUNNING FREQUENCY (f_o)

TYPICAL TRACKING RANGE AS A FUNCTION OF INPUT SIGNAL

CHANGE OF FREE RUNNING OSCILLATOR FREQUENCY AS A FUNCTION OF RANGE CONTROL CURRENT

CHANGE OF FREE RUNNING OSCILLATOR FREQUENCY AS A FUNCTION OF FINE TUNING CIRCUIT

FREE RUNNING OSCILLATOR FREQUENCY AS A FUNCTION OF VCO TIMING CAPACITANCE

NORMALIZED TRACKING RANGE AS A FUNCTION OF RANGE CONTROL CURRENT

Courtesy Signetics Corp.

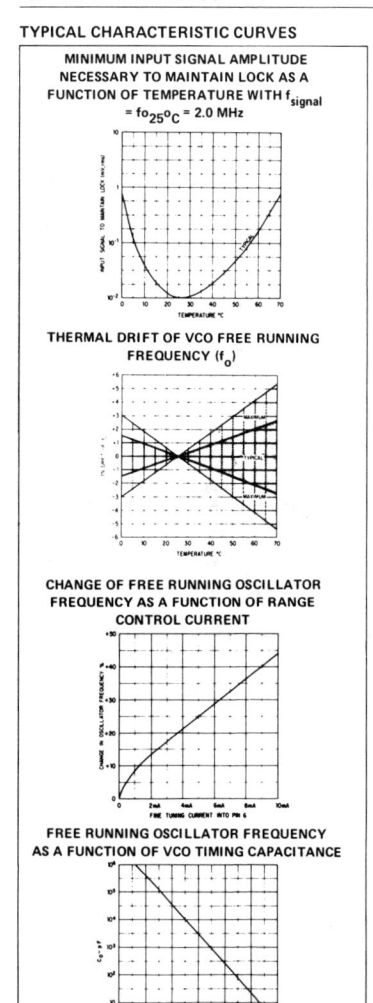

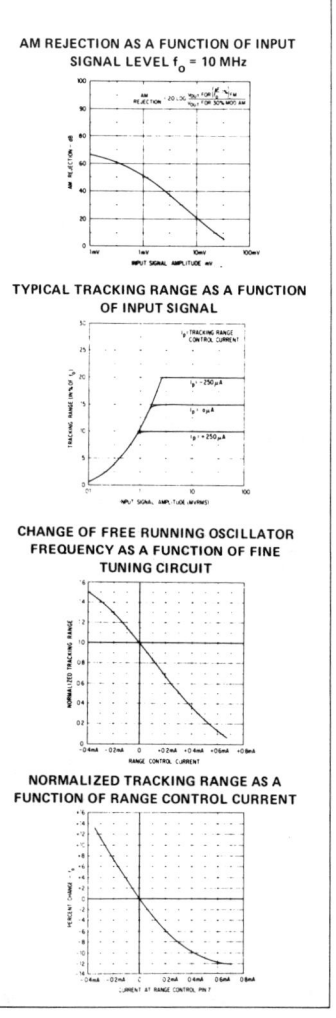

561 – PHASE LOCKED LOOP

EXTERNAL CONTROLS

1. **Loop Low Pass Filter (Pins 14 and 15)**
The equivalent circuit for the loop low-pass filter can be represented as:

where R_A (6kΩ) is the effective resistance seen looking into Pin #14 or Pin #15.
The corresponding filter transfer characteristics are:

$$\frac{V_2}{V_1}(S) = F(S) = \frac{1 + S R_1 C_1}{1 + S(R_1 + R_A) C_1}$$

where S is the complex frequency variable.

2. **Loop Gain (Threshold) Control**
The overall Phase Lock of loop gain can be reduced by connecting a feedback resistor, R_F, across the low-pass filter terminals, Pins #14 and #15. This causes the loop gain and the detection sensitivity to decrease by a factor ($\alpha < 1$), where

$$\alpha = \frac{R_F}{2R_A + R_F}$$

Reduction of loop gain may be desirable at high input signal levels ($V_{in} > 30$ mV) and at high frequencies ($f_o > 5$MHz) where excessively high PLL loop gain may cause instability within the loop.

3. **Tracking Range Control (Pin 7)**
Any bias current, I_P, injected into the tracking range control, reduces the tracking range of the PLL by decreasing the output of the limiter. The variation of the tracking range and the center frequency, as a function of I_P, are shown in the characteristic curves with I_P defined positive going into the tracking range control terminal. This terminal is normally at a DC level of +0.6 Volts and presents an impedance of 600Ω.

4. **External Fine Tuning (Pin 6)**
Any bias current injected into the fine tuning terminal increases the frequency of oscillation, f_o, as shown in the characteristic curves. This current is defined Positive into the fine tuning terminal. This terminal is at a typical DC level of +1.3 Volts and has a dynamic impedance of 100Ω to ground.

5. **Offset Adjustment (Pin 11)**
Application of a bias voltage to the offset adjustment terminal modifies the current in the output amplifier setting the DC level at the output. The effect on the loop is to modify the relationship between the VCO free running frequency and the lock range, allowing the VCO free running frequency to be positioned at different points throughout the lock range.
Nominally this terminal is at +4V DC and has an input impedance of 3kΩ. The offset adjustment is optional. The characteristics specified correspond to operation of the circuit with this terminal open circuited.

6. **De-emphasis Filter (Pin 10)**
The de-emphasis terminal is normally used when the PLL is used to demodulate Frequency Modulated Audio signals. In this application, a capacitor from this terminal to ground provides the required de-emphasis. For other applications, this terminal may be used for band shaping the output signal. The 3 dB bandwidth of the output amplifier in the system block diagram (see Figure 2 .) is related to the de-emphasis capacitor, C_D, as:

$$f_{3dB} = \frac{1}{2\pi R_D C_D}$$

where R_D is the 8000 ohm resistance seen looking into the de-emphasis terminal.
When the PLL system is utilized for signal conditioning, and the loop error voltage is not utilized, de-emphasis terminal should be AC grounded.

7. **AM Post-Detection Filter (Pin 1)**
The capacitor C_x connected between Pin #1 and ground serves as a low-pass filter for synchronous AM detection with a transfer characteristic, $F_2(S)$, given as:

$$F_2(S) = \frac{1}{1 + S R_x C_x}$$

where $R_x = 8$kΩ is the resistance seen looking into Pin #1.

Courtesy Signetics Corp.

PHASE LOCKED LOOP 562

LINEAR INTEGRATED CIRCUITS

DESCRIPTION

The NE562B Phase Locked Loop (PLL) is a monolithic signal conditioner and demodulator system, comprising a VCO, phase comparator, amplifier and low pass filter, interconnected as shown in the accompanying block diagram. The center frequency of the PLL is determined by the free running frequency (f_o) of the VCO. This VCO frequency is set by an external capacitor. The low pass filter, which determines the capture characteristics of the loop, is formed by two capacitors and two resistors at the phase comparator output.

This PLL has two sets of differential inputs, one for the FM/RF input and one for the phase comparator local oscillator input. Both sets of inputs can be used in either a differential or single-ended mode. The FM/RF inputs to the comparator are self-biased. An internally regulated voltage source is provided to bias the phase comparator local oscillator inputs. The VCO output, at high level and in differential form, is available for driving logic circuits in signal conditioning and synchronization, frequency multiplication and division applications. Terminals are also provided for the optional extension of the tracking range.

The monolithic signal conditioner-demodulator system is useful over a wide range of frequencies from less than 1 Hz to more than 15 MHz with an adjustable tracking range of ±1% to ±15%.

FEATURES
- FREQUENCY MULTIPLICATION AND DIVISION
- SIGNAL CONDITIONING AND SIDE-BAND SUPPRESSION
- FM DEMODULATION WITHOUT TUNED CIRCUITS
- NARROW BANDPASS – TO ±1%
- ADJUSTABLE TRACKING RANGE – TO ±15%
- EXACT FREQUENCY DUPLICATION IN HIGH NOISE ENVIRONMENT
- HIGH LINEARITY – 1% DISTORTION MAXIMUM AT 1% DEVIATION

APPLICATIONS
FREQUENCY SYNTHESIZERS
DATA SYNCHRONIZERS
SIGNAL CONDITIONING
TRACKING FILTERS
TELEMETRY DECODERS
MODEMS
FM IF STRIPS AND DEMODULATORS
TONE DECODERS
FSK RECEIVERS
WIDEBAND HIGH LINEARITY FM DEMODULATORS

PIN CONFIGURATION

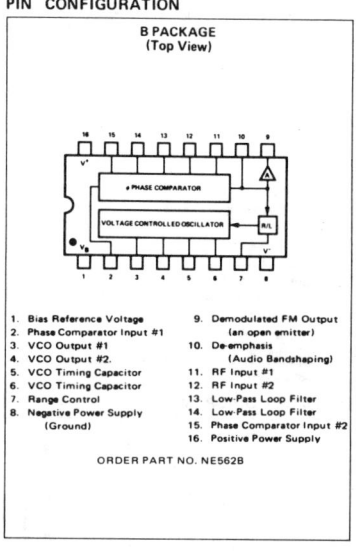

B PACKAGE
(Top View)

1. Bias Reference Voltage
2. Phase Comparator Input #1
3. VCO Output #1
4. VCO Output #2
5. VCO Timing Capacitor
6. VCO Timing Capacitor
7. Range Control
8. Negative Power Supply (Ground)
9. Demodulated FM Output (an open emitter)
10. De-emphasis (Audio Bandshaping)
11. RF Input #1
12. RF Input #2
13. Low-Pass Loop Filter
14. Low-Pass Loop Filter
15. Phase Comparator Input #2
16. Positive Power Supply

ORDER PART NO. NE562B

BLOCK DIAGRAM

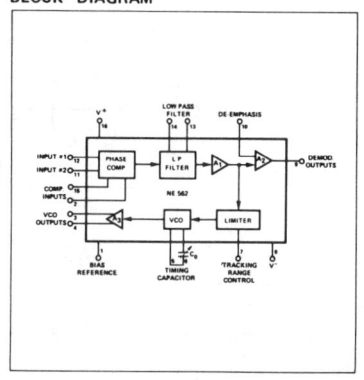

Courtesy Signetics Corp.

169

562 — PHASE LOCKED LOOP

ABSOLUTE MAXIMUM RATINGS (Limiting values above which serviceability may be impaired)

Maximum Operating Voltage	30V
Input Voltage	3V rms
Storage Temperature	$-65°C$ to $150°C$
Operating Temperature	$0°C$ to $70°C$
Power Dissipation	300mW

GENERAL ELECTRICAL CHARACTERISTICS

(15,000 ohms pin 9 to ground, 12,000 ohms pins 3 and 4 to ground, pins 2 and 15 to pin 1 through 1000 ohms, input to pin 11 or 12 with unused input at AC ground, range control not connected and V^+ = 18 volts unless otherwise specified. $T_A = 25°C$.)

CHARACTERISTICS	MIN	TYP	MAX	UNITS	TEST CONDITIONS
Lowest Practical Operating Frequency		0.1		Hz	
Maximum Operating Frequency	15	30		MHz	
Supply Current	10	12	14	mA	
Minimum Input Signal for Lock		200		µV	
Dynamic Range		80		dB	
VCO Temp Coefficient*		±0.06	±0.15	%/°C	Measured at 2 MHz
VCO Supply Voltage Regulation		±0.3	±2	%/V	Measured at 2 MHz
Input Resistance		2		kΩ	
Input Capacitance		4		pF	
Input DC Level	+12	+14	+16	V	
Output DC Level	+12	+14	+16	V	
Available Output Swing		4		V_{p-p}	Measured at Pin 9
AM Rejection*	30	40		dB	See Definition of Terms
De-emphasis Resistance		8		kΩ	
Bias Reference		+8		V	

*ACC Test Sub Group C.

SCHEMATIC DIAGRAM

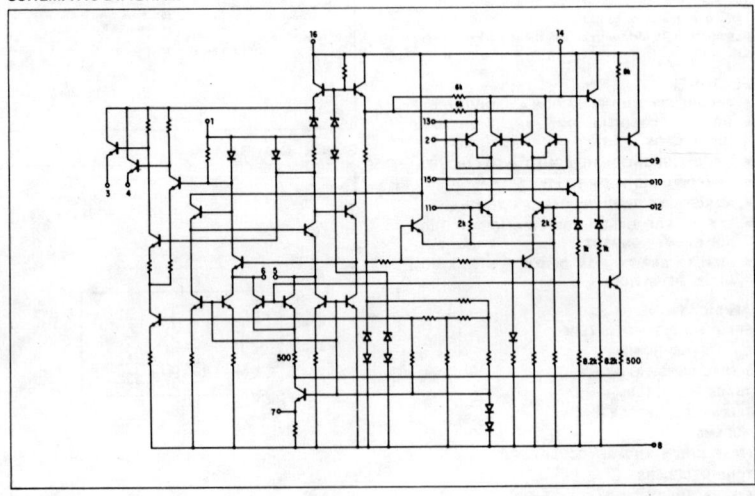

Courtesy Signetics Corp.

170

562 – PHASE LOCKED LOOP

TYPICAL CHARACTERISTIC CURVES

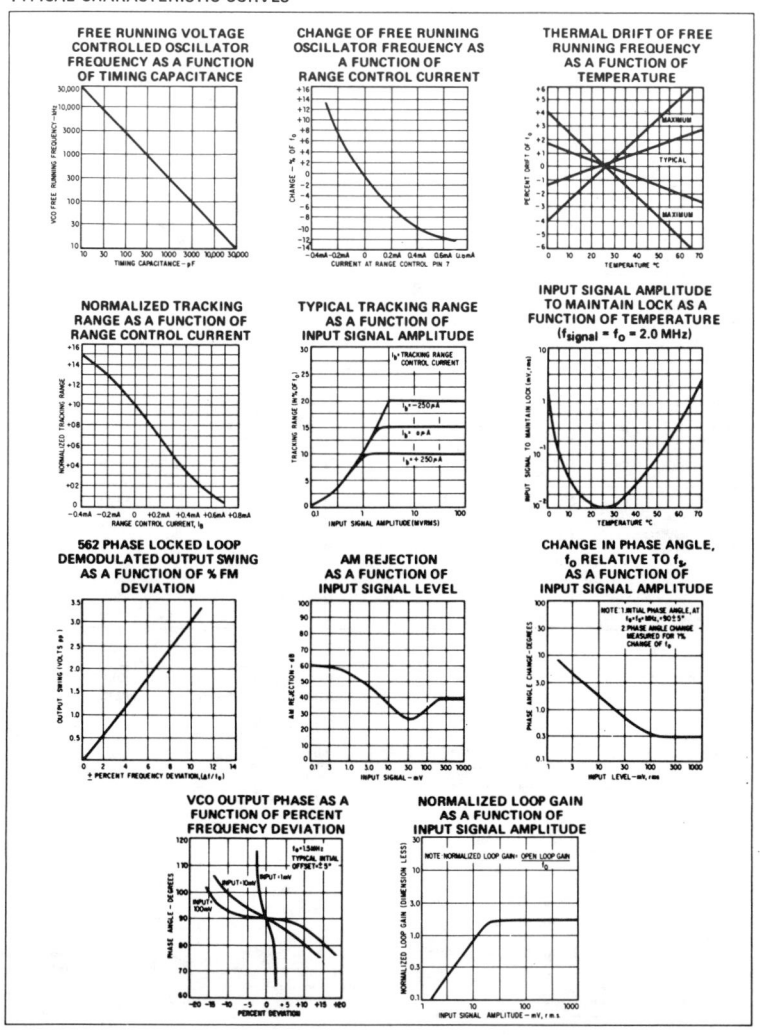

FREE RUNNING VOLTAGE CONTROLLED OSCILLATOR FREQUENCY AS A FUNCTION OF TIMING CAPACITANCE

CHANGE OF FREE RUNNING OSCILLATOR FREQUENCY AS A FUNCTION OF RANGE CONTROL CURRENT

THERMAL DRIFT OF FREE RUNNING FREQUENCY AS A FUNCTION OF TEMPERATURE

NORMALIZED TRACKING RANGE AS A FUNCTION OF RANGE CONTROL CURRENT

TYPICAL TRACKING RANGE AS A FUNCTION OF INPUT SIGNAL AMPLITUDE

INPUT SIGNAL AMPLITUDE TO MAINTAIN LOCK AS A FUNCTION OF TEMPERATURE ($f_{signal} = f_o = 2.0$ MHz)

562 PHASE LOCKED LOOP DEMODULATED OUTPUT SWING AS A FUNCTION OF % FM DEVIATION

AM REJECTION AS A FUNCTION OF INPUT SIGNAL LEVEL

CHANGE IN PHASE ANGLE, f_o RELATIVE TO f_s, AS A FUNCTION OF INPUT SIGNAL AMPLITUDE

VCO OUTPUT PHASE AS A FUNCTION OF PERCENT FREQUENCY DEVIATION

NORMALIZED LOOP GAIN AS A FUNCTION OF INPUT SIGNAL AMPLITUDE

Courtesy Signetics Corp.

562 – PHASE LOCKED LOOP

ELECTRICAL CHARACTERISTICS FOR FM APPLICATIONS (15,000 ohms pin 9 to ground, input to pin 11 or pin 12, AC ground unused input, range control not connected and $V^+ = 18$ volts. $T_A = 25°C$.)

CHARACTERISTICS	LIMITS MIN	LIMITS TYP	LIMITS MAX	UNITS	TEST CONDITIONS
10.7 MHz Operation	Deviation 75 kHz Source Impedance = 50Ω				
Detection Threshold		200	500	µV	
Demodulated Output Amplitude	30	70		mV rms	V_{in} = 1 mV rms Modulation Frequency 1 kHz
Distortion*		0.5		% T.H.D.	V_{in} = 1 mV rms Modulation Frequency 1 kHz
Signal to Noise Ratio $\frac{S+N}{N}$		35		dB	V_{in} = 1 mV rms Modulation Frequency 1 kHz
4.5 MHz Operation	Deviation = 25 kHz, Source Impedance = 50Ω				
Detection Threshold		200	500	µV rms	
Demodulated Output Amplitude	30	60		mV rms	V_{in} = 1 mV rms Modulation Frequency 1 kHz
Distortion		0.5		% T.H.D.	V_{in} = 1 mV rms Modulation Frequency 1 kHz
Signal to Noise Ratio $\frac{S+N}{N}$		35		dB	V_{in} = 1 mV rms Modulation Frequency 1 kHz
Wide Deviation	$\Delta F/f_o$ = 5% Input = 4.5 MHz Deviation = 225 kHz @ 1 kHz Modulation Rate				
Detection Threshold		1	5	mV	
Demodulated Output		1		V rms	V_{in} = 5 mV rms
Distortion	0.3	0.8		% T.H.D.	V_{in} = 5 mV rms
Signal to Noise Ratio $\frac{S+N}{N}$		50		dB	V_{in} = 5 mV rms

*ACC Test Sub Group C.

ELECTRICAL CHARACTERISTICS FOR SIGNAL CONDITIONER AND FREQUENCY SYNTHESIS APPLICATIONS (Input to pin 11 or pin 12, AC ground unused input, range control not connected, V^+ = 18 volts. T_A = 25°C.)

CHARACTERISTIC	LIMITS MIN	LIMITS TYP	LIMITS MAX	UNITS	TEST CONDITIONS
Tracking Range	±5	±15		% of f_o	200 mV p-p square wave input
Input Resistance		2		kΩ	
Input Capacitance		4		pF	
Input DC Level		4		V	
VCO Output Impedance		1.3	2.5	kΩ	
VCO Output Swing	3	4.5		V p-p	
VCO Output DC Level		12		V	
VCO Signal/Noise Ratio		60		dB	Inputs at AC ground

TEST CIRCUIT

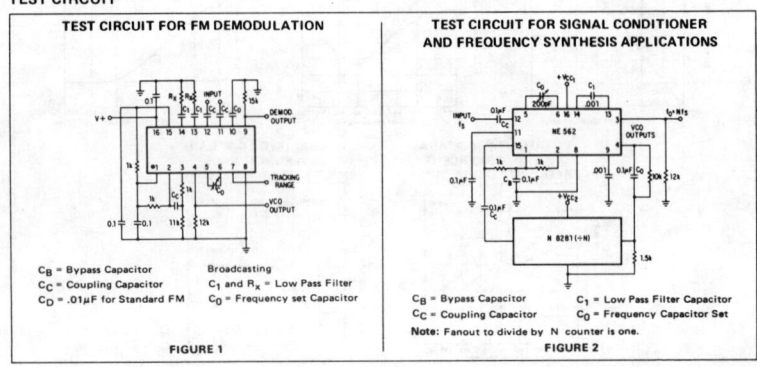

TEST CIRCUIT FOR FM DEMODULATION

C_B = Bypass Capacitor
C_C = Coupling Capacitor
C_D = .01µF for Standard FM
Broadcasting
C_1 and R_x = Low Pass Filter
C_0 = Frequency set Capacitor

FIGURE 1

TEST CIRCUIT FOR SIGNAL CONDITIONER AND FREQUENCY SYNTHESIS APPLICATIONS

C_B = Bypass Capacitor
C_C = Coupling Capacitor
C_1 = Low Pass Filter Capacitor
C_0 = Frequency Capacitor Set

Note: Fanout to divide by N counter is one.

FIGURE 2

Courtesy Signetics Corp.

562 − PHASE LOCKED LOOP

562 APPLICATIONS INFORMATION
1. BIAS REFERENCE
Pin 1 of the 562 is an internally regulated bias reference voltage supply which should be used as a source of bias current for the phase comparator input terminals, Pins 2 and 15. Biasing may be achieved as shown in Figure 3.

FIGURE 3

2. PHASE COMPARATOR LOOP INPUTS
Of the Signetics high frequency phase locked loops, the 562 is unique in that the loop is open between the VCO and the phase comparator. Once biasing of the comparator is accomplished, as described in Bias Reference above, loop closure can be accomplished by capacitive coupling between either one or both inputs of the phase comparator and the VCO output. A divider or counter may be enclosed in the loop at this point for frequency synthesis applications or a flip-flop may be used to ensure that the output waveform has a 50% duty cycle. If large signal swings, greater than 2 volts, are to be applied to the phase comparator inputs, a 1000 ohm current limiting buffer resistor should be used in series with the coupling capacitors.

3. VCO OUTPUT
Square wave VCO outputs of both polarities (0°C and 180°C) buffered by an amplifier are available at pins 3 and 4. For proper operation of the buffer amplifier, pins 3 and 4 must be returned to ground (or the negative supply) through resistors, typically 12,000 ohms. The value of these resistors may be reduced provided that total power dissipated in the 562 does not exceed 300 milliwatts or the total average current in each emitter does not exceed 4 mA. The output amplitude is typically 4.5 volts peak referenced at +12 volts with respect to pin 8.

4. VCO TUNING
Setting the free-running frequency of the VCO is accomplished easily with one timing capacitor connected between pins 5 and 6. For the 562 Phase Locked Loop, fine tuning of the free-running frequency may be accomplished in either or both of two ways. The first method uses a trimmer capacitor connected in parallel with the VCO timing capacitor. This is the simplest technique and requires the smallest number of extra components but at the lower frequencies may be difficult to implement. The second technique incorporates two resistors and a voltage source. The resistors are connected between each of the timing capacitor terminals and a voltage source as shown in Figure 4.

FIGURE 4

The percent change in the VCO free-running frequency, f_o, as a function of the voltage applied to point (A) is shown in the curves of Figure 5. Note that with this fine tuning technique, it is possible to *increase* the VCO free-running frequency to a value greater than possible with just a trimmer capacitor alone. A formula for the approximation of the VCO frequency as a function of the voltage at point (A), the resistance values and the starting frequency, is given below:

$$f = f_o \left[1 - \frac{V_A - 6.4}{1300R} \right]$$

The recommended resistance range of R is 20,000 to 60,000 ohms.

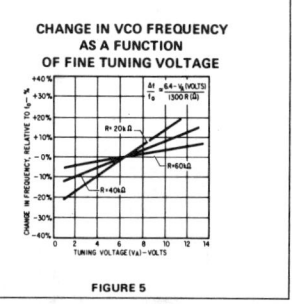

CHANGE IN VCO FREQUENCY AS A FUNCTION OF FINE TUNING VOLTAGE

FIGURE 5

5. LOOP GAIN CHARACTERISTICS
The overall open loop gain of the 562 PLL can be expressed as:
$$K_0 = K_1 K_2$$
where:

K_0 = total open loop gain

K_1 = phase comparator and amplifier conversion gain

K_2 = VCO conversion gain

The VCO conversion gain, K_2, is the change of VCO frequency per unit of error voltage. In this particular design, it is numerically equal to the VCO frequency, i.e.,

$$K_2 = f_o \text{ Hz/Volt}$$
or
$$K_2 = 2\pi f_o \text{ radians/Volt-second}$$

The phase comparator and amplifier conversion gain, K_1, is proportional to input signal amplitude for low input levels, $V_s \leqslant 40\text{mV rms}$, and it is constant and equal to about 1.5 volts/radian for higher input amplitudes. Therefore, K_1 can be approximated as:

$$K_1 \cong \frac{.04 \, V_s}{\sqrt{1 + \left(\frac{V_s}{40}\right)^2}}$$

where

V_s = input signal in mV rms.

Courtesy Signetics Corp.

562 − PHASE LOCKED LOOP

562 APPLICATIONS INFORMATION (Cont'd.)

6. SIGNAL INPUT
The input structure is basically differential and may be used in this manner. Biasing is supplied to the input terminals from an internal regulated supply so signal inputs must be capacitively coupled. In most applications where the input is single-ended, the unused input should be bypassed to ground.

7. DEMODULATED OUTPUT
Pin 9 is a low impedance output terminal for the loop error voltage. It is at this point that the demodulated FM output is obtained. When used, it must be biased by a resistor to ground (or negative supply), and the resistor value may be adjusted downward provided that the output current does not exceed 5mA or the dissipation in the 562 does not exceed the absolute maximum ratings. When not used, pin 9 may be left open.

8. DE-EMPHASIS FILTER
The de-emphasis terminal, pin 10, is normally required when the PLL is used to demodulate Frequency Modulated Audio signals. In this application, a capacitor from this terminal to ground provides the required de-emphasis. For other applications it may be used to shape the output response. The 3 dB bandwidth of the output amplifier is related to the de-emphasis capacitor, C_D, as:

$$f_{3dB} = \frac{1}{2\pi R_D C_D}$$

where R_D is 8000 ohms.

When the PLL system is utilized for applications not requiring the use of the output amplifier, pin 10 should be by-passed to ground.

9. TRACKING RANGE CONTROL (Pin 7)
Any bias current, I_p, injected into the tracking range control, reduces the tracking range of the PLL by decreasing the output of the limiter. The variation of the tracking range and the center frequency, as a function of I_p, are shown in the characteristic curves with I_p defined positive going into the tracking range control terminal. This terminal is normally at a DC level of +0.6 volts and presents an impedance of 600Ω.

10. LOW-PASS FILTER
In most applications, a loop low-pass filter should be connected between pins 13 and 14 and ground. It is used to set the loop response time, controlling the capture range and the rejection of out of band information. Four filter configurations and their transfer functions are shown in Figures 6 through 9. For VCO operating frequencies below 5 MHz, configurations shown in Figures 6 and 7 may be used. At higher frequencies, configurations shown in Figures 8 and 9 should be used to ensure loop stability. R is the impedance seen looking into the low pass filter terminals, Pins 13 and 14; and, in the 562, is nominally 6000 ohms.

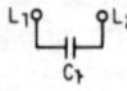

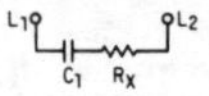

FIGURE 6   FIGURE 7

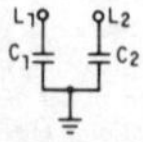

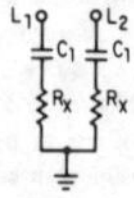

FIGURE 8   FIGURE 9

11. LOOP GAIN (Threshold) CONTROL
The overall Phase Locked Loop gain can be reduced by connecting a resistor, R_F, across the low-pass filter terminals, pins 13 and 14. This causes the loop gain and the detection sensitivity to decrease by a factor α, where:

$$\alpha = \frac{R_F}{12,000 + R_F}$$

Reduction of loop gain may be desirable at operating frequencies greater than 5 MHz because, at these frequencies, high loop gain may cause instability.

12. STATIC LOOP PHASE-ERROR
When the PLL is in lock, the VCO outputs have a nominal ±90°C phase shift with respect to the input signal. Due to internal offsets, this nominal angle at perfect lock condition may shift a few degrees, typically ±5°C or less.

Courtesy Signetics Corp.

PHASE LOCKED LOOP 564

DESCRIPTION
The NE564 is a versatile, high frequency Phase Locked Loop designed for operation up to 50MHz. As shown in the block diagram, the NE564 consists of a VCO, limiter, phase comparator, and post detection processor.

APPLICATIONS
- High speed modems
- FSK receivers and transmitters
- Frequency synthesizers
- Signal generators

FEATURES
- Operation with single 5V supply
- TTL compatible inputs and outputs
- Operation to 50MHz
- Operates as a modulator
- External loop gain control
- Reduced carrier feedthrough
- No elaborate filtering needed in FSK applications

PIN CONFIGURATION

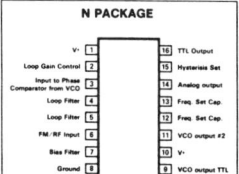

N PACKAGE

Pin	Function
1	V+
2	Loop Gain Control
3	Input to Phase Comparator from VCO
4	Loop Filter
5	Loop Filter
6	FM/RF Input
7	Bias Filter
8	Ground
9	VCO output TTL
10	V-
11	VCO output #2
12	Freq. Set Cap.
13	Freq. Set Cap.
14	Analog output
15	Hysteresis Set
16	TTL Output

ABSOLUTE MAXIMUM RATINGS

PARAMETER		RATING	UNIT
V+	Supply voltage		V
	Pin 1	14	
	Pin 10	6	
P_D	Power dissipation	400	mW
T_A	Operating temperature	0 to 70	°C
t_{stg}	Storage temperature	-65 to 150	°C

TYPICAL PERFORMANCE CHARACTERISTICS

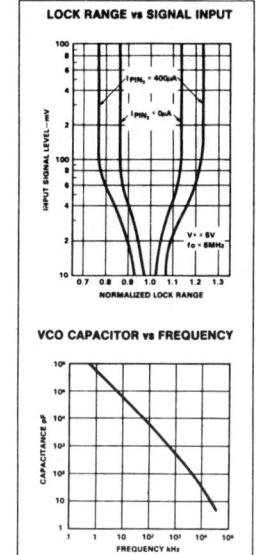

LOCK RANGE vs SIGNAL INPUT

VCO CAPACITOR vs FREQUENCY

BLOCK DIAGRAM

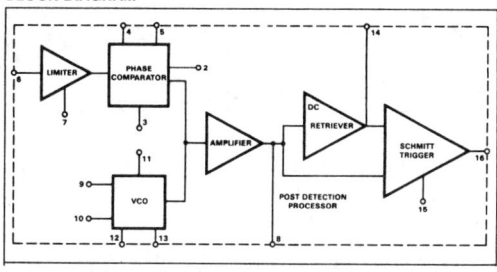

FUNCTIONAL DESCRIPTION
The NE564 is a monolithic phase locked loop with a post detection processor. The use of Schottky clamped transistors and optimized device geometries extends the frequency of operation to 50MHz. In addition to the classical PLL applications, the NE564 can be used as a modulator with a controllable frequency deviation.

The output voltage of the PLL can be written as shown in the following equation:

Equation 1

$$v_O = \frac{(f_{in} - f_O)}{K_{VCO}}$$

K_{VCO} = conversion gain of the VCO
f_{in} = frequency of the input signal
f_O = free running frequency of the VCO

The process of recovering FSK signals involves the conversion of the PLL output into digital, logic compatible signals. For high data rates, a considerable amount of carrier

Courtesy Signetics Corp.

175

PHASE LOCKED LOOP 564

564 • N

Design Formula
Free running frequency of VCO is shown by the following equation:

Equation 4

$$f_O = \frac{1}{16 R_C C_1} \text{ in Hz}$$

$R_C = 100\,\Omega$
C_1 = external cap in farads

The loop filter diagram shown is explained by the following equation:

Equation 5

$$F(s) = \frac{1}{1 + sRC_2}$$

$R = R_{12} = R_{13} = 1.3\,k\Omega$

LOOP FILTER

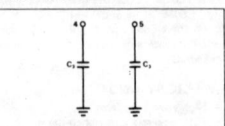

EQUIVALENT SCHEMATIC

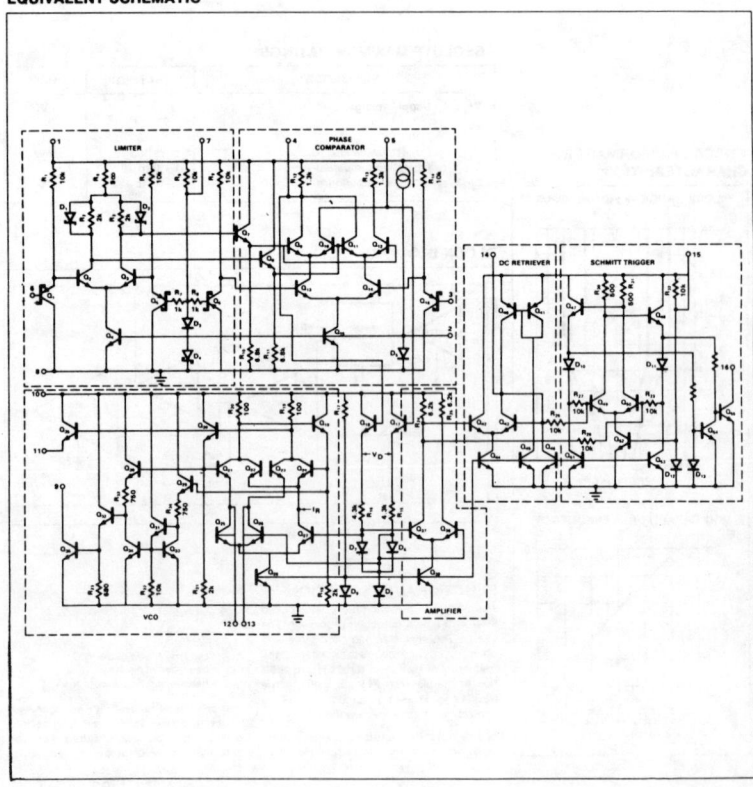

Courtesy Signetics Corp.

PHASE LOCKED LOOP 564

DC ELECTRICAL CHARACTERISTICS V+ = 5V, T$_A$ = 25°C unless otherwise specified

	PARAMETER	TEST CONDITIONS	Min	Typ	Max	UNIT
f$_O$	Lock range	T$_A$ = 25°C, I$_2$ = 400µA	25	40		%
	Frequency of operation of VCO		45	50		MHz
	Frequency drift with temperature	T$_A$ = 0°C to 70°C, f$_O$ = 5MHz		400	850	ppm/°C
	Frequency change with supply voltage	V+ = 4.5V to 5.5V		3	6	%/V
	Demodulated output voltage	1% input deviation	10	14		mV
		10% input deviation, f$_O$ = 5MHz	100	140		
	Output voltage linearity			3		%
	Signal to noise ratio			40		dB
	AM rejection			35		dB
I$_{CC}$	Supply current	5V		30	40	mA
I$_{LC}$	Leakage current	Pin 9		1	10	µA
	Output current	Pin 9			6	mA
V+	Supply voltage	Pin 1	4.5		12	V
		Pin 10	4.5		5.5	

will be present at the output due to the the use of complicated filters, a comparator with hysterisis or Schmitt trigger is required. With the conversion gain of the VCO fixed, the output voltage as given by Equation 1 varies according to the frequency deviation of f$_{in}$ from f$_O$. Since this differs from system to system, it is necessary that the hysterisis of the Schmitt trigger be capable of being changed, so that it can be optimized for a particular system. This is accomplished in the 564 by varying the voltage at pin 15 which results in a change of the hysterisis of the Schmitt trigger.

For FSK signals, an important factor to be considered is the drift in the free running frequency of the VCO itself. If this changes due to temperature, according to Equation 1 it will lead to a change in the dc levels of the PLL output, and consequently to errors in the digital output signal. This is especially true for narrow band signals where the deviation in f$_{in}$ itself may be less than the change in f$_O$ due to temperature. This effect can be eliminated if the dc or average value of the signal is retrieved and used as the reference to the comparator. In this manner, variations in the dc levels of the PLL output do not affect the FSK output.

VCO Section

Due to its inherent high frequency performance, an emitter coupled oscillator is used in the VCO. In the circuit, shown in the equivalent schematic, transistors Q$_{21}$ and Q$_{23}$ with current sources Q$_{25}$-Q$_{26}$ form the basic oscillator. The free running frequency of the oscillator is shown in the following equation:

Equation 2

$$f_O = \frac{1}{16 R_C C_1}$$

R$_C$ = R$_{19}$ = R$_{20}$
C$_1$ = frequency setting external capacitor

Variation of V$_d$ changes the frequency of the oscillator. As indicated by Equation 2, the frequency of the oscillator has a negative temperature coefficient due to the positive temperature coefficient of the monolithic resistor. To compensate for this, a current I$_R$ with negative temperature coefficient is introduced to achieve a low frequency drift with temperature.

Phase Comparator Section

The phase comparator consists of a double balanced modulator with a limiter amplifier to improve AM rejection. Schottky clamped vertical PNPs are used to obtain TTL level inputs. The loop gain can be varied changing the current in Q$_4$ and Q$_{15}$ which effectively changes the gain of the differential amplifiers. This can be accomplished by introducing a current at pin 2.

Post Detection Processor Section

The post detection processor consists of a unity gain transconductance amplifier and comparator. The amplifier can be used as a dc retriever for demodulation of FSK signals, and as a post detection filter for linear FM demodulation. The comparator has adjustable hysterisis so that phase jitter in the output signal can be eliminated.

As shown in the equivalent schematic, the dc retriever is formed by the transconductance amplifier Q$_{42}$-Q$_{43}$ at the output (pin 14). This forms an integrator whose output voltage is shown in the following equation:

Equation 3

$$V_o = \frac{g_m}{C_2} \int V_{in} \, dt$$

g$_m$ = transconductance of the amplifier
C$_2$ = capacitor at the output (pin 14)
V$_{in}$ = signal voltage at amplifier input

With proper selection of C$_2$, the integrator time constant can be varied so that the output voltage is the dc or average value of the input signal for use in FSK, or as a post detection filter in linear demodulation.

The comparator with hysterisis is made up of Q$_{49}$-Q$_{50}$ with positive feedback being provided by Q$_{47}$-Q$_{48}$. The hysterisis is varied by changing the current in Q$_{52}$ with a resulting variation in the loop gain of the comparator. This method of hysterisis control, which is a dc control, provides symmetric variation around the nominal value.

Courtesy Signetics Corp.

PHASE LOCKED LOOP 564

ANALOG APPLICATIONS MEMO

FM DEMODULATOR

The NE564 can be used as an FM demodulator. The connections for operation at 5V and 12V are shown in Figures 1 and 2 respectively. The input signal is ac coupled with the output signal being extracted at pin 14. Loop filtering is provided by the capacitors at pins 4 and 5 with additional filtering being provided by the capacitor at pin 14. Since the conversion gain of the VCO is not very high, to obtain sufficient demodulated output signal the frequency deviation in the input signal should be fairly high (1% or higher).

FM DEMODULATOR WITH TTL COMPATIBLE OUTPUT SIGNAL

An FM demodulator with the output signal being a TTL signal can be obtained from the NE564 by connecting it as shown in Figure 3. This operation requires the use of the dc retriever, the capacitance for which is connected at pin 14. The hysterisis of the Schmitt trigger can be adjusted by connecting a potentiometer at pin 15. The output signal appears at pin 16, which requires an external resistor. If necessary, the duty cycle of the output signal can be adjusted by applying a voltage at pin 14 (around 2.5V) and varying it. The connection for a similar application appears in Figure 4.

GATED PLL DEMODULATOR

The lock range adjust pin of the NE564 can be used to gate the PLL when it is operating in the demodulator mode. The circuit is connected as shown in Figure 5. The gating voltage which can be a TTL signal is applied to pin 2. When this voltage is high, the loop is in lock and the demodulated output signal appears at pin 16. When the input to pin 2 is low, the loop is out of lock and the VCO will be at its center frequency. It is also possible to use pin 2 to adjust the loop gain so that a large capture range and small lock range can be obtained.

MODULATION TECHNIQUES

The NE564 phase locked loop can be modulated at either the loop filter ports (pins 4 and 5) or the input port (pin 6) as shown in Figure 6. The approximate modulation frequency can be determined from the frequency conversion gain curve shown in Figure 7. This curve will be appropriate for signals injected into pins 4 and 5.

FREQUENCY SYNTHESIS

Frequency multiplication can be achieved with the NE564 with the insertion of a counter (digital frequency divider) in the loop.

A block diagram is shown in Figure 8 and the associated performance characteristic

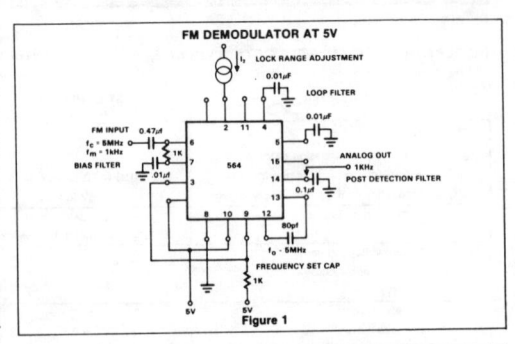

Figure 1

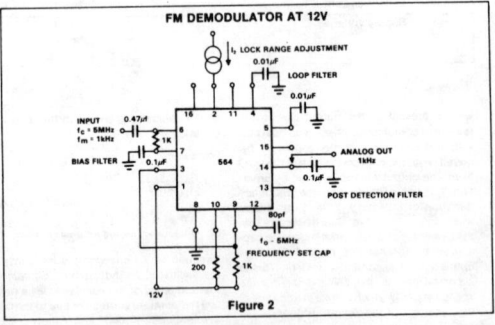

Figure 2

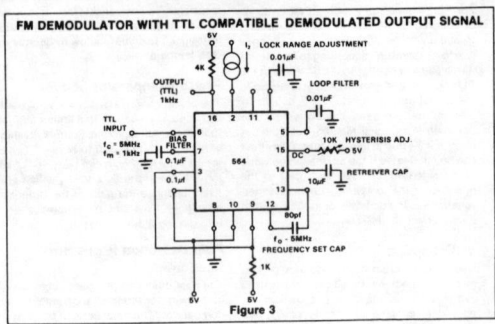

Figure 3

Courtesy Signetics Corp.

PHASE LOCKED LOOP 564

ANALOG APPLICATIONS MEMO

curve in Figure 7. Here the loop is broken between the VCO and the phase comparator and a counter is inserted. In this case, the fundamental of the divided VCO frequency is locked to the input frequency so that the VCO is actually running at a multiple of the input frequency. The amount of multiplication is determined by the counter. An obvious practical application of this multiplication property is the use of the NE564 in wide range frequency synthesizers.

In frequency multiplication applications it is important to take into account that the phase comparator is actually a mixer and that its output contains sum and difference frequency components. The difference frequency component is dc and is the error voltage which drives the VCO to keep the NE564 in lock. The sum frequency components (of which the fundamental is twice the frequency of the input signal) if not well filtered, will induce incidental FM on the VCO output. This occurs because the VCO is running at many times the frequency of the input signal and the sum frequency component which appears on the control voltage to the VCO causes a periodic variation of its frequency about the desired multiple. For frequency multiplication it is generally necessary to filter quite heavily to remove this sum frequency component. The tradeoff, of course, is a reduced capture range and a more underdamped loop transient response.

Figure 4 — 12V DEMODULATOR WITH DIGITAL OUTPUT

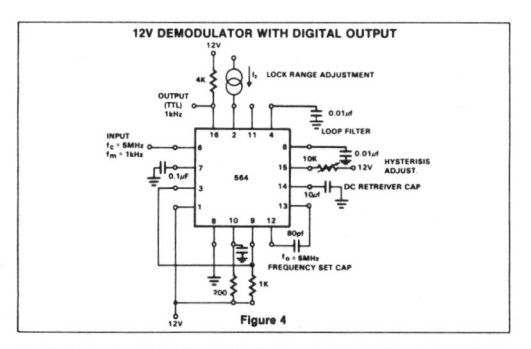

Figure 5 — GATED PLL DEMODULATOR

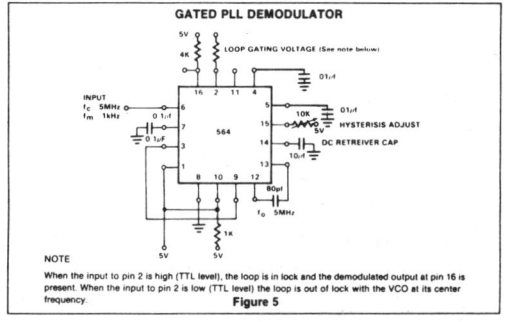

NOTE
When the input to pin 2 is high (TTL level), the loop is in lock and the demodulated output at pin 16 is present. When the input to pin 2 is low (TTL level) the loop is out of lock with the VCO at its center frequency.

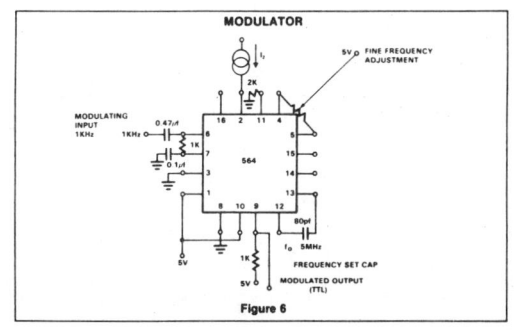

Figure 6 — MODULATOR

Courtesy Signetics Corp.

179

PHASE LOCKED LOOP 564

ANALOG APPLICATIONS MEMO

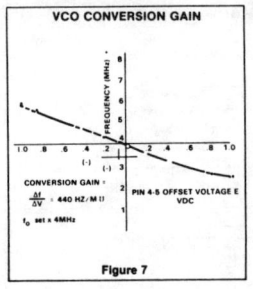

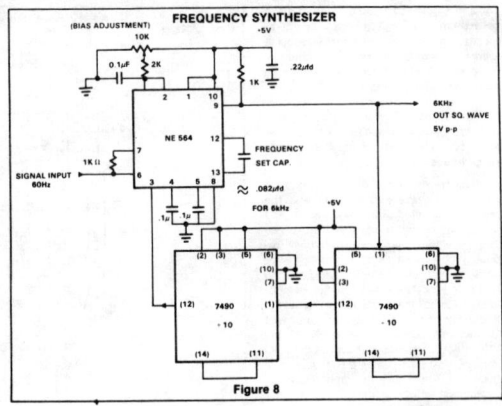

Figure 7. VCO CONVERSION GAIN

Figure 8. FREQUENCY SYNTHESIZER

Courtesy Signetics Corp.

PHASE LOCKED LOOP 565

LINEAR INTEGRATED CIRCUITS

DESCRIPTION

The SE/NE565 Phase-Locked Loop (PLL) is a self-contained, adaptable filter and demodulator for the frequency range from 0.001 Hz to 500 kHz. The circuit comprises a voltage-controlled oscillator of exceptional stability and linearity, a phase comparator, an amplifier and a low-pass filter as shown in the block diagram. The center frequency of the PLL is determined by the free-running frequency of the VCO; this frequency can be adjusted externally with a resistor or a capacitor. The low-pass filter, which determines the capture characteristics of the loop, is formed by an internal resistor and an external capacitor.

FEATURES

- EXTREME STABILITY OF CENTER FREQUENCY (200ppm/°C typ)
- WIDE RANGE OF OPERATING VOLTAGE (±5 to ±12 VOLTS) WITH VERY SMALL FREQUENCY DRIFT (100ppm/% typ)
- VERY HIGH LINEARITY OF DEMODULATED OUTPUT (0.2% typ)
- CENTER FREQUENCY PROGRAMMING BY MEANS OF A RESISTOR, CAPACITOR, VOLTAGE OR CURRENT
- TTL AND DTL COMPATIBLE SQUARE-WAVE OUTPUT; LOOP CAN BE OPENED TO INSERT DIGITAL FREQUENCY DIVIDER
- HIGHLY LINEAR TRIANGLE WAVE OUTPUT
- REFERENCE OUTPUT FOR CONNECTION OF COMPARATOR IN FREQUENCY DISCRIMINATOR
- BANDPASS, ADJUSTABLE FROM <±1% to >±60%
- FREQUENCY ADJUSTABLE OVER 10 TO 1 RANGE WITH SAME CAPACITOR

BLOCK DIAGRAM

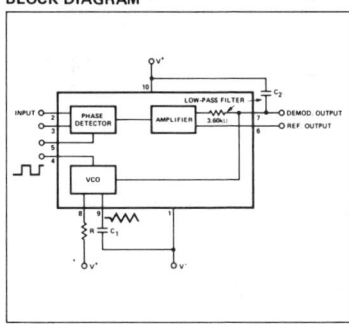

PIN CONFIGURATIONS

A PACKAGE (Top View)

1. V^-
2. Input
3. Input
4. VCO Output
5. Phase Comparator VCO Input
6. Reference Output
7. Demodulated Output
8. External R for VCO
9. External C for VCO
10. V^+
11. NC
12. NC
13. NC
14. NC

ORDER PART NOS. SE565A/NE565A

K PACKAGE

1. V^-
2. Input
3. Input
4. VCO Output
5. Phase Comparator VCO Input
6. Reference Output
7. Demodulated Output
8. External R for VCO
9. External C for VCO
10. V^+

ORDER PART NOS. SE565K/NE565K

APPLICATIONS

FREQUENCY SHIFT KEYING
MODEMS
TELEMETRY RECEIVERS
TONE DECODERS
SCA RECEIVERS
WIDEBAND FM DISCRIMINATORS
DATA SYNCHRONIZERS
TRACKING FILTERS
SIGNAL RESTORATION
FREQUENCY MULTIPLICATION & DIVISION

Courtesy Signetics Corp.

181

SE/NE565 — PHASE LOCKED LOOP

ABSOLUTE MAXIMUM RATINGS (limiting values above which serviceability may be impaired)

Maximum Operating Voltage	26V
Storage Temperature	$-65°C$ to $150°C$
Power Dissipation	300mW

ELECTRICAL CHARACTERISTICS ($T_A = 25°C$, $V_{CC} = \pm 6$ Volts unless otherwise noted)

PARAMETER	TEST CONDITIONS	SE565 MIN	SE565 TYP	SE565 MAX	NE565 MIN	NE565 TYP	NE565 MAX	UNITS
SUPPLY REQUIREMENTS								
Supply Voltage		±5		±12	±5		±12	V
Supply Current			8	12.5		8	12.5	mA
INPUT CHARACTERISTICS								
Input Impedance	$-4V \leqslant V_2, V_3 \leqslant +1V$	7	10		5	10		kΩ
Input Level Required for Tracking	$f_0 = 50$ kHz, ±10% frequency deviation	10	1		10	1		mVrms
VCO CHARACTERISTICS								
Center Frequency Maximum Value	$C_1 = 2.7$ pF	300	500			500		kHz
Distribution	Distribution taken about $f_0 \simeq 50$ kHz $R_1 = 5.0$k, $C_1 = 1200$ pF	-10	0	+10	-30	0	+30	%
Drift with Temperature	$f_0 = 50$ kHz	+75	+100	+525		+200		ppm/°C
Drift with Supply Voltage	$f_0 = 50$ kHz, $V_{CC} = \pm 6$ to ± 7 Volts		0.1	1.0		0.2	1.5	%/V
Triangle Wave								
Output Voltage Level			0			0		
Amplitude		2	2.4	3	2	2.4	3	Vp-p
Linearity			0.2			0.5		%
Square Wave								
Logical "1" Output Voltage	$f_0 = 50$ kHz, $V_{CC} = \pm 6$ Volts	+4.9	+5.2		+4.9	+5.2		V
Logical "0" Output Voltage	$f_0 = 50$ kHz, $V_{CC} = \pm 6$ Volts		-0.2	+0.2		-0.2	+0.2	V
Duty Cycle	$f_0 = 50$ kHz	45	50	55	40	50	60	%
Rise Time			20	100		20		nsec
Fall Time			50	200		50		nsec
Output Current (sink)		0.6	1		0.6	1		mA
Output Current (source)		5	10		5	10		mA
DEMODULATED OUTPUT CHARACTERISTICS								
Output Voltage Level	(pin 7) $V_{CC} = \pm 6$ Volts	4.25	4.5	4.75	4.0	4.5	5.0	V
Maximum Voltage Swing	(pin 7)		2			2		Vp-p
Output Voltage Swing	±10% frequency deviation	250	300		200	300		mVp-p
Total Harmonic Distortion			0.2	0.75		0.2	1.5	%
Output Impedance			3.6			3.6		kΩ
Offset Voltage $|V_6 - V_7|$	$T_A = 25°C$		30	100		50	200	mV
vs Temperature (drift)			50			100		$\mu V/°C$
AM Rejection		30	40			40		dB

NOTES:
1. Both input terminals (pins 2 and 3) must receive identical dc bias. This bias may range from 0 volts to -4 volts.
2. The external resistance for frequency adjustment (R_1) must have a value between 2kΩ and 20kΩ.
3. Output voltage swings negative as input frequency increases.
4. Output not buffered.

Courtesy Signetics Corp.

SE/NE565 — PHASE LOCKED LOOP

TYPICAL PERFORMANCE CHARACTERISTICS

POWER SUPPLY CURRENT AS A FUNCTION OF SUPPLY VOLTAGE

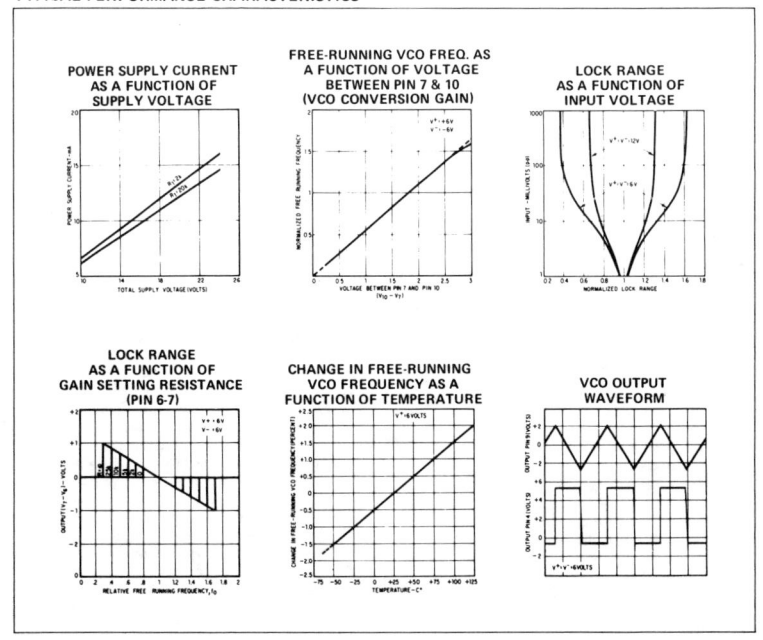

FREE-RUNNING VCO FREQ. AS A FUNCTION OF VOLTAGE BETWEEN PIN 7 & 10 (VCO CONVERSION GAIN)

LOCK RANGE AS A FUNCTION OF INPUT VOLTAGE

LOCK RANGE AS A FUNCTION OF GAIN SETTING RESISTANCE (PIN 6-7)

CHANGE IN FREE-RUNNING VCO FREQUENCY AS A FUNCTION OF TEMPERATURE

VCO OUTPUT WAVEFORM

SCHEMATIC DIAGRAM

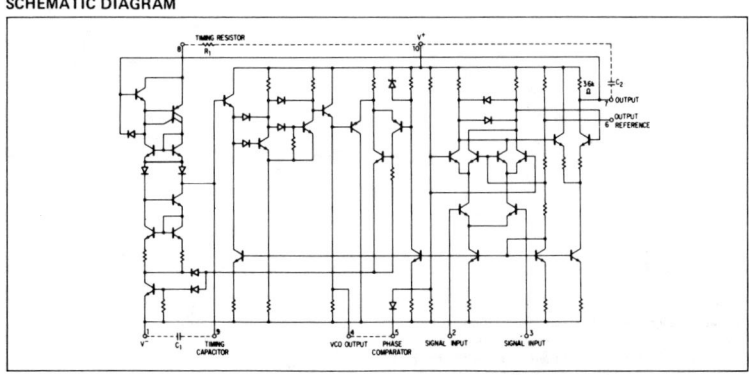

Courtesy Signetics Corp.

SE/NE565 – PHASE LOCKED LOOP

DESIGN FORMULAS

Free running frequency of VCO $f_o = \dfrac{1.2}{4R_1C_1}$ in Hz

Lock range $f_L = \pm \dfrac{8 f_o}{V_{CC}}$ in Hz

Capture range $f_C \simeq \pm \dfrac{1}{2\pi}\sqrt{\dfrac{2\pi f_L}{\tau}}$

where $\tau = (3.6 \times 10^3) \times C_2$

DEFINITION OF TERMS

FREE-RUNNING FREQUENCY (f_o)
Frequency of VCO without input signal, both inputs grounded.

CAPTURE-RANGE
That range of frequencies about f_o over which the loop will acquire lock with an input signal initially starting out of lock.

LOCK-RANGE OR TRACKING-RANGE
That range of frequencies in the vicinity of f_o over which the VCO, once locked to the input signal, will remain locked.

TYPICAL APPLICATIONS

FM DEMODULATION
The 565 Phase Locked Loop is a general purpose circuit designed for highly-linear FM demodulation. During lock, the average dc level of the phase comparator output signal is directly proportional to the frequency of the input signal. As the input frequency shifts, it is this output signal which causes the VCO to shift its frequency to match that of the input. Consequently, the linearity of the phase comparator output with frequency is determined by the voltage-to-frequency transfer function of the VCO.

Because of its unique and highly linear VCO, the 565 PLL can lock to and track an input signal over a very wide range (typically ±60%) with very high linearity (typically, within 0.5%).

A typical connection diagram is shown in Figure 1. The VCO free-running frequency is given approximately by $f_o = \dfrac{1.2}{4R_1C_1}$ and should be adjusted to be at the center of the input signal frequency range. C_1 can be any value, but R_1 should be within the range of 2000 to 20,000 ohms with an optimum value on the order of 4000 ohms. The source can be direct coupled if the dc resistances seen from pins 2 and 3 are equal and there is no dc voltage difference between the pins. A short between pins 4 and 5 connects the VCO to the phase comparator. Pin 6 provides a dc reference voltage that is close to the dc potential of the demodulated output (pin 7). Thus, if a resistance (R_2 in Figure 1) is connected between pins 6 and 7, the gain of the output stage can be reduced with little change in the dc voltage level at the output. This allows the lock range to be decreased with little change in the free-running frequency. In this manner the lock range can be decreased from ±60% of f_o to approximately ±20% of f_o (at ±6V).

A small capacitor (typically 0.001 µF) should be connected between pins 7 and 8 to eliminate possible oscillation in the control current source.

A single-pole loop filter is formed by the capacitor C_2, connected between pin 7 and positive supply, and an internal resistance of approximately 3600 ohms.

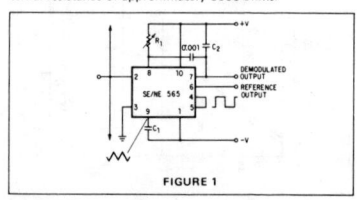

FIGURE 1

FREQUENCY SHIFT KEYING (FSK)

FSK refers to data transmission by means of carrier which is shifted between two preset frequencies. This frequency shift is usually accomplished by driving a VCO with the binary data signal so that the two resulting frequencies correspond to the "0" and "1" states (commonly called space and mark) of the binary data signal.

A simple scheme using the 565 to receive FSK signals of 1070 Hz and 1270 Hz is shown in Figure 2. As the signal appears at the input, the loop locks to the input frequency and tracks it between the two frequencies with a corresponding dc shift at the output.

The loop filter capacitor C_2 is chosen smaller than usual to eliminate overshoot on the output pulse, and a three-stage RC ladder filter is used to remove the carrier component from the output. The band edge of the ladder filter is chosen to be approximately half way between the maximum keying rate (in this case 300 baud or 150 Hz) and twice the input frequency (approximately 2200 Hz). The output signal can now be made logic compatible by connecting a voltage comparator between the output and pin 6 of the loop. The free-running frequency is adjusted with R_1 so as to result in a slightly-positive voltage at the output at f_{in} = 1070 Hz.

The input connection is typical for cases where a dc voltage is present at the source and therefore a direct connection is not desirable. Both input terminals are returned to ground with identical resistors (in this case, the values are chosen to effect a 600-ohm input impedance).

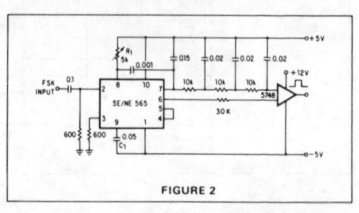

FIGURE 2

Courtesy Signetics Corp.

SE/NE565 — PHASE LOCKED LOOP

FREQUENCY MULTIPLICATION
There are two methods by which frequency multiplication can be achieved using the 565:
1. Locking to a harmonic of the input signal.
2. Inclusion of a digital frequency divider or counter in the loop between the VCO and phase comparator.

The first method is the simplest, and can be achieved by setting the free-running frequency of the VCO to a multiple of the input frequency. A limitation of this scheme is that the lock range decreases as successively higher and weaker harmonics are used for locking. If the input frequency is to be constant with little tracking required, the loop can generally be locked to any one of the first 5 harmonics. For higher orders of multiplication, or for cases where a large lock range is desired, the second scheme is more desirable. An example of this might be a case where the input signal varies over a wide frequency range and a large multiple of the input frequency is required.

A block diagram of the second scheme is shown in Figure 3. Here the loop is broken between the VCO and the phase comparator, and a frequency divider is inserted. The fundamental of the divided VCO frequency is locked to the input frequency in this case, so that the VCO is actually running at a multiple of the input frequency. The amount of multiplication is determined by the frequency divider. A typical connection scheme is shown in Figure 4. To set up the circuit, the frequency limits of the input signal must be determined. The free-running frequency of the VCO is then adjusted by means of R_1 and C_1 (as discussed under FM demodulation) so that the output frequency of the divider is midway between the input frequency limits. The filter capacitor, C_2, should be large enough to eliminate variations in the demodulated output voltage (at pin 7), in order to stabilize the VCO frequency. The output can now be taken as the VCO squarewave output, and its fundamental will be the desired multiple of the input frequency (f_1) as long as the loop is in lock.

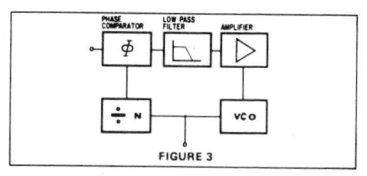

FIGURE 3

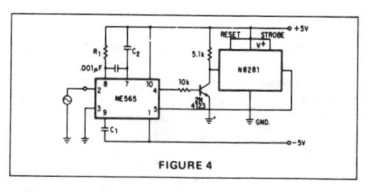

FIGURE 4

SCA (BACKGROUND MUSIC) DECODER
Some FM stations are authorized by the FCC to broadcast uninterrupted background music for commercial use. To do this a frequency modulated subcarrier of 67 kHz is used. The frequency is chosen so as not to interfere with the normal stereo or monaural program; in addition, the level of the subcarrier is only 10% of the amplitude of the combined signal.

The SCA signal can be filtered out and demodulated with the NE565 Phase Locked Loop without the use of any resonant circuits. A connection diagram is shown in Figure 5. This circuit also serves as an example of operation from a single power supply.

A resistive voltage divider is used to establish a bias voltage for the input (pins 2 and 3). The demodulated (multiplex) FM signal is fed to the input through a two-stage high-pass filter, both to effect capacitive coupling and to attenuate the strong signal of the regular channel. A total signal amplitude, between 80 mV and 300 mV, is required at the input. Its source should have an impedance of less than 10,000 ohms.

The Phase Locked Loop is tuned to 67 kHz with a 5000 ohm potentiometer; only approximate tuning is required, since the loop will seek the signal.

The demodulated output (pin 7) passes through a three-stage low-pass filter to provide de-emphasis and attenuate the high-frequency noise which often accompanies SCA transmission. Note that no capacitor is provided directly at pin 7; thus, the circuit is operating as a first-order loop. The demodulated output signal is in the order of 50 mV and the frequency response extends to 7 kHz.

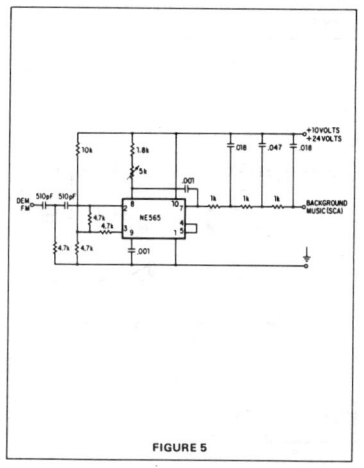

FIGURE 5

Courtesy Signetics Corp.

185

TONE DECODER PHASE LOCKED LOOP | 567

LINEAR INTEGRATED CIRCUITS

DESCRIPTION
The SE/NE 567 tone and frequency decoder is a highly stable phase-locked loop with synchronous AM lock detection and power output circuitry. Its primary function is to drive a load whenever a sustained frequency within its detection band is present at the self-biased input. The bandwidth center frequency, and output delay are independently determined by means of four external components.

FEATURES
- WIDE FREQUENCY RANGE (.01Hz TO 500kHz)
- HIGH STABILITY OF CENTER FREQUENCY
- INDEPENDENTLY CONTROLLABLE BANDWIDTH (0 TO 14 PERCENT)
- HIGH OUT-BAND SIGNAL AND NOISE REJECTION
- LOGIC-COMPATIBLE OUTPUT WITH 100mA CURRENT SINKING CAPABILITY
- INHERENT IMMUNITY TO FALSE SIGNALS
- FREQUENCY ADJUSTMENT OVER A 20 TO 1 RANGE WITH AN EXTERNAL RESISTOR

APPLICATIONS
TOUCH TONE® DECODING
CARRIER CURRENT REMOTE CONTROLS
ULTRASONIC CONTROLS (REMOTE TV, ETC.)
COMMUNICATIONS PAGING
FREQUENCY MONITORING AND CONTROL
WIRELESS INTERCOM
PRECISION OSCILLATOR

BLOCK DIAGRAM

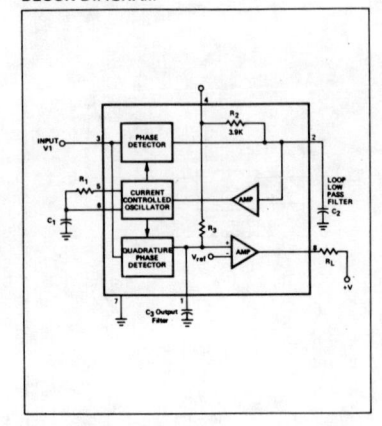

PIN CONFIGURATION

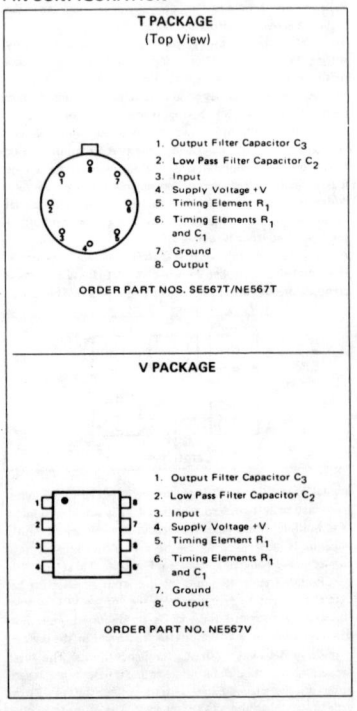

T PACKAGE
(Top View)

1. Output Filter Capacitor C_3
2. Low Pass Filter Capacitor C_2
3. Input
4. Supply Voltage +V
5. Timing Element R_1
6. Timing Elements R_1 and C_1
7. Ground
8. Output

ORDER PART NOS. SE567T/NE567T

V PACKAGE

1. Output Filter Capacitor C_3
2. Low Pass Filter Capacitor C_2
3. Input
4. Supply Voltage +V
5. Timing Element R_1
6. Timing Elements R_1 and C_1
7. Ground
8. Output

ORDER PART NO. NE567V

ABSOLUTE MAXIMUM RATINGS:

Operating Temperature	0°C to 70°C NE567
	-55°C to 125°C SE567
Operating Voltage	10V
Positive Voltage at Input	0.5V above Supply Voltage (Pin 4)
Negative Voltage at Input	-10 VDC
Output Voltage (collector of output transistor)	15 VDC
Storage Temperature	-65°C to 150°C
Power Dissipation	300mW

Courtesy Signetics Corp.

567 — TONE DECODER PHASE LOCKED LOOP

ELECTRICAL CHARACTERISTICS (V+ = 5.0 Volts, T_A = 25°C unless noted)

CHARACTERISTICS	SE567 MIN	SE567 TYP	SE567 MAX	NE567 MIN	NE567 TYP	NE567 MAX	UNITS	TEST CONDITIONS
CENTER FREQUENCY (NOTE 1)								
Highest Center Frequency (f_o)	100	500		100	500		kHz	
Center Frequency Stability (Note 2)		35±140			35±140		ppm/°C	−55 to 125°C
		35±60			35±60		ppm/°C	0 to 70°C
Center Frequency Shift with Supply Voltage		0.5	1		0.7	2	%/Volt	f_o = 100KHz
DETECTION BANDWIDTH								
Largest Detection Bandwidth	12	14	16	10	14	18	% of f_o	f_o = 100KHz
Largest Detection Bandwidth Skew		1	2		2	3	% of f_o	
Largest Detection Bandwidth – Variation with Temperature		±0.1			±0.1		%/°C	V_i = 300mVrms
Largest Detection Bandwidth – Variation with Supply Voltage		±2			±2		%/Volt	V_i = 300mVrms
INPUT								
Input Resistance		20			20		KΩ	
Smallest Detectable Input Voltage (V_i)		20	25		20	25	mV rms	I_L = 100mA, f_i = f_o
Largest No-Output Input Voltage	10	15		10	15		mV rms	I_L = 100mA, f_i = f_o
Greatest Simultaneous Outband Signal to Inband Signal Ratio		+6			+6		dB	
Minimum Input Signal to Wideband Noise Ratio		−6			−6		dB	Bn = 140KHz
OUTPUT								
Fastest On-Off Cycling Rate		f_o/20			f_o/20			
"1" Output Leakage Current		0.01	25		0.01	25	μA	
"0" Output Voltage		0.2	0.4		0.2	0.4	Volt	I_L = 30mA
		0.6	1.0		0.6	1.0	Volt	I_L = 100mA
Output Fall Time (Note 3)		30			30		n sec	R_L = 50Ω
Output Rise Time (Note 3)		150			150		n sec	R_L = 50Ω
GENERAL								
Operating Voltage Range	4.75		9.0	4.75		9.0	Volts	
Supply Current – Quiescent		6	8		7	10	mA	
Supply Current – Activated		11	13		12	15	mA	R_L = 20KΩ
Quiescent Power Dissipation		30			35		mW	

NOTES:
1. Frequency determining resistor R_1 should be between 1 and 20KΩ.
2. Applicable over 4.75 to 5.75 volts. See graphs for more detailed information.
3. Pin 8 to Pin 1 feedback R_L network selected to eliminate pulsing during turn-on and turn-off.

Courtesy Signetics Corp.

567 – TONE DECODER PHASE LOCKED LOOP

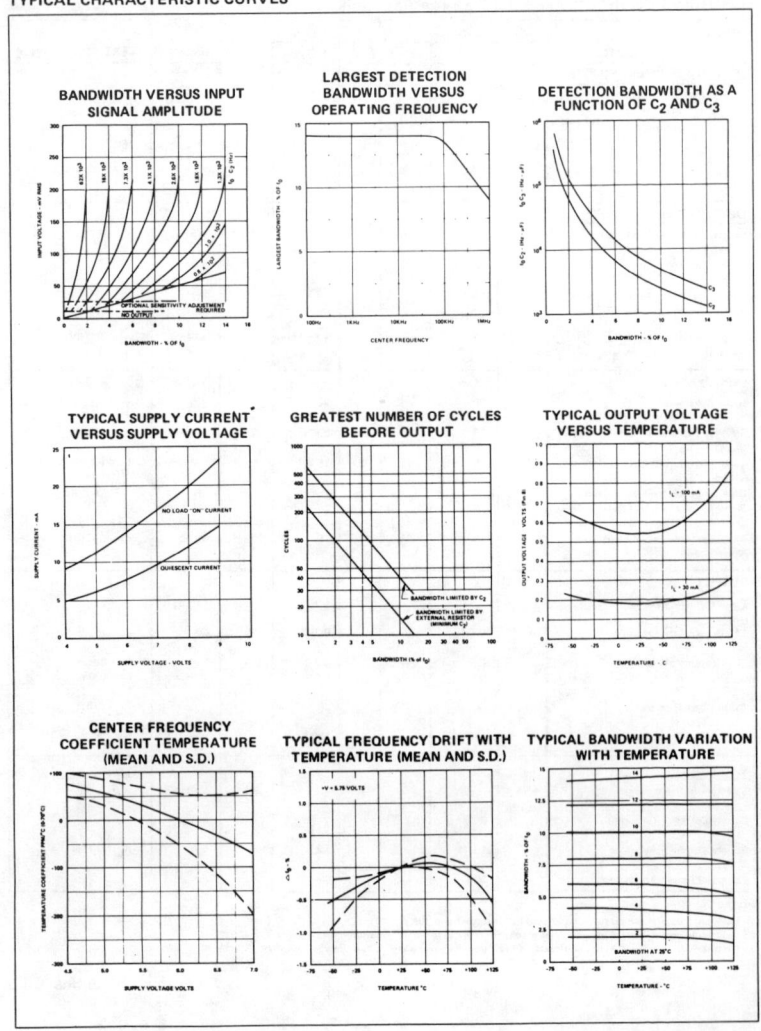

TYPICAL CHARACTERISTIC CURVES

Courtesy Signetics Corp.

567 – TONE DECODER PHASE LOCKED LOOP

TYPICAL CHARACTERISTIC CURVES (Cont'd.)

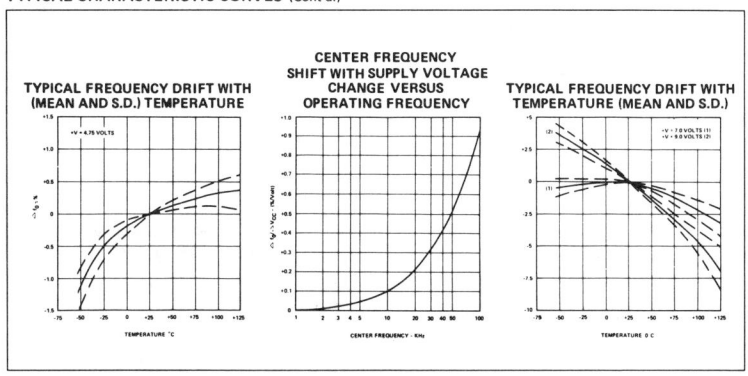

TYPICAL FREQUENCY DRIFT WITH (MEAN AND S.D.) TEMPERATURE

CENTER FREQUENCY SHIFT WITH SUPPLY VOLTAGE CHANGE VERSUS OPERATING FREQUENCY

TYPICAL FREQUENCY DRIFT WITH TEMPERATURE (MEAN AND S.D.)

SCHEMATIC DIAGRAM

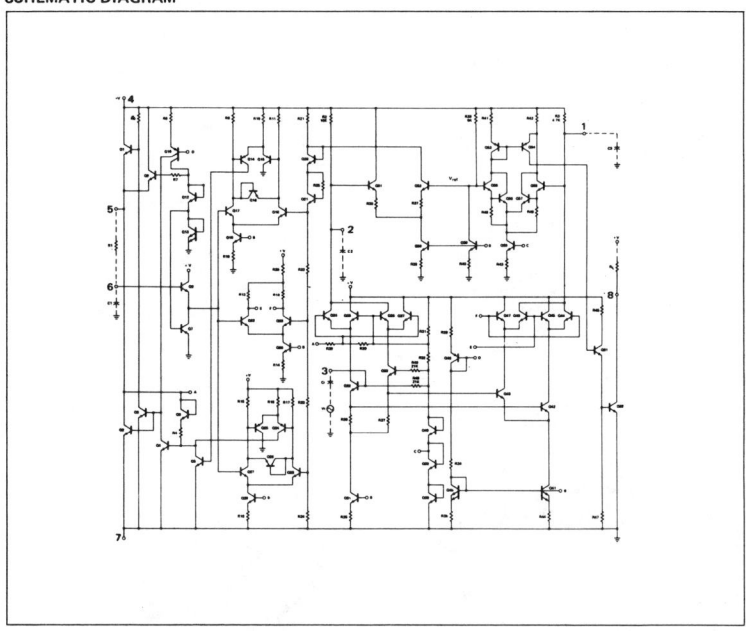

Courtesy Signetics Corp.

567 – TONE DECODER PHASE LOCKED LOOP

DESIGN FORMULAS

$$f_0 \simeq \frac{1.1}{R_1 C_1}$$

$$BW \simeq 1070 \sqrt{\frac{V_i}{f_0 C_2}} \text{ in \% of } f_0, V: \lesssim 200 \text{ nV}$$

Where
V_i = Input Voltage (Volts)
C_2 = Low-Pass Filter Capacitor (μF)

PHASE LOCKED LOOP TERMINOLOGY

CENTER FREQUENCY (f_0)
The free-running frequency of the current controlled oscillator (CCO) in the absence of an input signal.

DETECTION BANDWIDTH (BW)
The frequency range, centered about f_0, within which an input signal above the threshold voltage (typically 20mV rms) will cause a logical zero state on the output. The detection bandwidth corresponds to the loop capture range.

LARGEST DETECTION BANDWIDTH
The largest frequency range within which an input signal above the threshold voltage will cause a logical zero state on the output. The maximum detection bandwidth corresponds to the loop lock range.

DETECTION BAND SKEW
A measure of how well the largest detection band is centered about the center frequency, f_0. The skew is defined as $(f_{max} + f_{min} - 2f_0)/f_0$ where f_{max} and f_{min} are the frequencies corresponding to the edges of the detection band. The skew can be reduced to zero if necessary by means of an optional centering adjustment.

TYPICAL RESPONSE

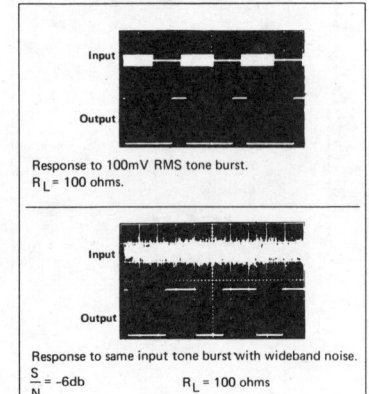

Response to 100mV RMS tone burst.
R_L = 100 ohms.

Response to same input tone burst with wideband noise.
$\frac{S}{N}$ = -6db R_L = 100 ohms
Noise Bandwidth = 140 Hz

OPERATING INSTRUCTIONS
Figure 1 shows a typical connection diagram for the 567. For most applications, the following three-step procedure will be sufficient for choosing the external components R_1, C_1, C_2 and C_3.

1. Select R_1 and C_1 for the desired center frequency. For best temperature stability, R_1 should be between 2K and 20K ohm, and the $R_1 C_1$ product should have sufficient stability, over the projected temperature range to meet the necessary requirements.

2. Select the low-pass capacitor, C_2, by referring to the Bandwidth versus Input Signal Amplitude graph. If the input amplitude variation is known, the appropriate value of $f_0 C_2$ necessary to give the desired bandwidth may be found. Conversely, an area of operation may be selected on this graph and the input level and C_2 may be adjusted accordingly. For example, constant bandwidth operation requires that input amplitude be above 200mVrms. The bandwidth, as noted on the graph, is then controlled solely by the $f_0 C_2$ product (F_0 (Hz), C_2 (pFd)).

3. The value of C_3 is generally non-critical. C_3 sets the band edge of a low pass filter which attenuates frequencies outside the detection band to eliminate spurious outputs. If C_3 is too small, frequencies just outside the detection band will switch the output stage on and off at the beat frequency, or the output may pulse on and off during the turn-on transient. If C_3 is too large, turn-on and turn-off of the output stage will be delayed until the voltage on C_3 passes the threshold voltage. (Such a delay may be desirable to avoid spurious outputs due to transient frequencies.) A typical minimum value for C_3 is $2C_2$.

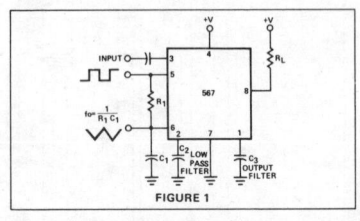

FIGURE 1

AVAILABLE OUTPUTS (Figure 2)
The primary output is the uncommitted output transistor collector, pin 8. When an in-band input signal is present, this transistor saturates; its collector voltage being less than 1.0 volt (typically 0.6V) at full output current (100mA). The voltage at pin 2 is the phase detector output, a linear function of frequency, over the range of 0.95 to 1.05 f_0, with a slope of about 20mV/% frequency deviation. The average voltage at pin 1 is, during lock, a function of the in-band input amplitude in accordance with the transfer characteristic given. Pin 5 is the controlled oscillator square wave output of magnitude $(V^+ - 2Vbe) \approx (V^+ - 1.4V)$ having a dc average of $V^+/2$. A 1KΩ load may be driven from pin 5. Pin 6 is an exponential triangle of 1 volt peak-to-peak

Courtesy Signetics Corp.

567 – TONE DECODER PHASE LOCKED LOOP

AVAILABLE OUTPUTS (Cont'd.)
with an average dc level of $V^+/2$. Only high impedance loads may be connected to pin 6 without affecting the CCO duty cycle or temperature stability.

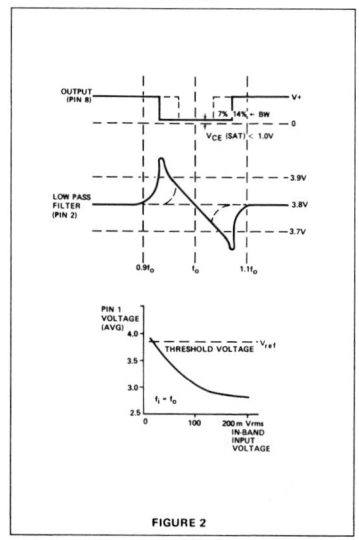

FIGURE 2

OPERATING PRECAUTIONS
A brief review of the following precautions will help the user attain the high level of performance of which the 567 is capable.

1. Operation in the high input level mode (above 200mV) will free the user from bandwidth variations due to changes in the in-band signal amplitude. The input stage is now limiting, however, so that out-band signals or high noise levels can cause an apparent bandwidth reduction as the in band signal is suppressed. Also, the limiting action will create in-band components from sub-harmonic signals, so the 567 becomes sensitive to signals at $f_o/3$, $f_o/5$, etc.

2. The 567 will lock onto signals near $(2n+1)f_o$ and will give an output for signals near $(4n+1)f_o$ where $n = 0, 1, 2,$ etc. Thus, signals at 5 f_o and 9 f_o can cause an unwanted output. If such signals are anticipated, they should be attenuated before reaching the 567 input.

3. Maximum immunity from noise and out-band signals is afforded in the low input level (Below 200mVrms) and reduced bandwidth operating mode. However, decreased loop damping causes the worse-case lock-up time to increase, as shown by the Greatest Number of Cycles Before Output vs. Bandwidth graph.

4. Due to the high switching speeds (20ns) associated with 567 operation, care should be taken in lead routing. Lead lengths should be kept to a minimum. The power supply should be adequately bypassed close to the 567 with an 0.01µF or greater capacitor; grounding paths should be carefully chosen to avoid ground loops and unwanted voltage variations. Another factor which must be considered is the effect of load energization on the power supply. For example, an incandescent lamp typically draws 10 times rated current at turn-on. This can cause supply voltage fluctuations which could, for example, shift the detection band of narrow-band systems sufficiently to cause momentary loss of lock. The result is a low-frequency oscillation into and out of lock. Such effects can be prevented by supplying heavy load currents from a separate supply, or increasing the supply filter capacitor.

SPEED OF OPERATION
Minimum lock-up time is related to the natural frequency of the loop. The lower it is, the longer becomes the turn-on transient. Thus, maximum operating speed is obtained when C_2 is at a minimum. When the signal is first applied, the phase may be such as to initially drive the controlled oscillator *away* from the incoming frequency rather than toward it. Under this condition, which is of course unpredictable, the lock-up transient is at its worst and the theoretical minimum lock-up time is not achievable. We must simply wait for the transient to die out.

The following expressions give the values of C_2 and C_3 which allow highest operating speeds for various band center frequencies. The minimum rate at which digital information may be detected without information loss due to the turn-on transient or output chatter is about 10 cycles per bit, corresponding to an information transfer rate of $f_o/10$ baud.

$$C_2 = \frac{130}{f_o} \mu F$$

$$C_3 = \frac{260}{f_o} \mu F$$

in cases where turn-off time can be sacrificed to achieve fast turn-on, the optional sensitivity adjustment circuit can be used to move the quiescent C_3 voltage lower (closer to the threshold voltage). However, sensitivity to beat frequencies, noise and extraneous signals will be increased.

OPTIONAL CONTROLS
The 567 has been designed so that, for most applications, no external adjustments are required. Certain applications, however, will be greatly facilitated if full advantage is taken of the added control possibilities available through the use of additional external components. In the diagrams given, typical values are suggested where applicable. For best results resistors used, except where noted, should have the same temperature coefficient. Ideally, silicon diodes would be low-resistivity types, such as forward-biased low-voltage zeners or forward-biased transistor base-emitter junctions. However, ordinary low-voltage diodes should be adequate for most applications.

Courtesy Signetics Corp.

191

567 – TONE DECODER PHASE LOCKED LOOP

SENSITIVITY ADJUSTMENT

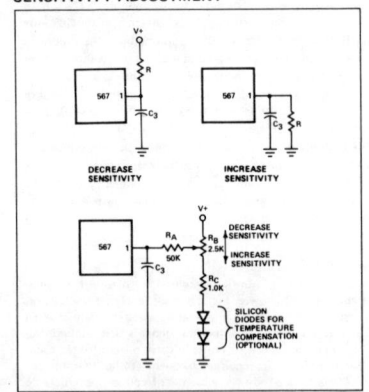

When operated as a very narrow band detector (less than 8 percent), both C_2 and C_3 are made quite large in order to improve noise and outband signal rejection. This will inevitably slow the response time. If, however, the output stage is biased closer to the threshold level, the turn-on time can be improved. This is accomplished by drawing additional current to terminal 1. Under this condition, the 567 will also give an output for lower-level signals (10m or lower).

By adding current to terminal 1, the output stage is biased further away from the threshold voltage. This is most useful when, to obtain maximum operating speed, C_2 and C_3 are made very small. Normally, frequencies just outside the detection band could cause false outputs under this condition. By desensitizing the output stage, the outband beat notes do not feed through to the output stage. Since the input level must be somewhat greater when the output stage is made less sensitive, rejection of third harmonics or in-band harmonics (of lower frequency signals) is also improved.

CHATTER PREVENTION

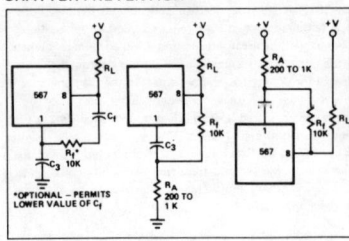

Chatter occurs in the output stage when C_3 is relatively small, so that the lock transient and the AC components at the quadrature phase detector (lock detector) output cause the output stage to move through its threshold more than once. Many loads, for example lamps and relays, will not respond to the chatter. However, logic may recognize the chatter as a series of outputs. By feeding the output stage output back to its input, (pin 1) the chatter can be eliminated. Three schemes for doing this are given above. All operate by feeding the first output step (either on or off) back to the input, pushing the input past the threshold until the transient conditions are over. It is only necessary to assure that the feedback time constant is not so large as to prevent operation at the highest anticipated speed. Although chatter can always be eliminated by making C_3 large, the feedback circuit will enable faster operation of the 567 by allowing C_3 to be kept small. Note that if the feedback time constant is made quite large, a short burst at the input frequency can be stretched into a long output pulse. This may be useful to drive, for example, stepping relays.

DETECTION BAND CENTERING (OR SKEW) ADJUSTMENT

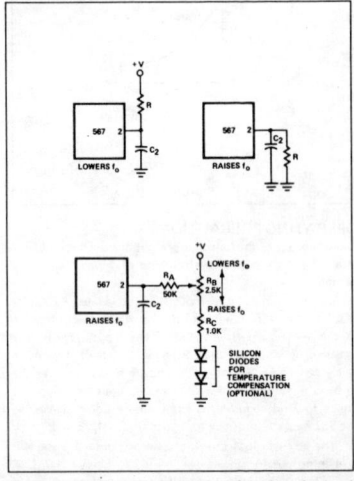

When it is desired to alter the location of the detection band (corresponding to the loop capture range) within the largest detection band (lock range), the circuits shown above can be used. By moving the detection band to one edge of the range, for example, input signal variations will expand the detection band in only one direction. This may prove useful when a strong but undesirable signal is expected on one side or the other of the center frequency. Since R_B also alters the duty cycle slightly, this method may be used to obtain a precise duty cycle when the 567 is used as an oscillator.

Courtesy Signetics Corp.

567 — TONE DECODER PHASE LOCKED LOOP

ALTERNATE METHOD OF BANDWIDTH REDUCTION

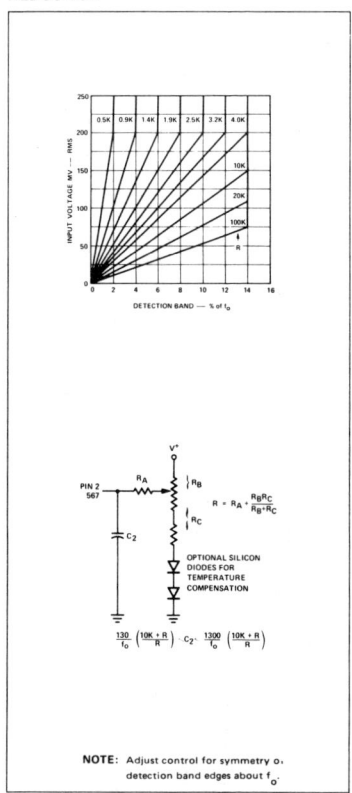

Although a large value of C_2 will reduce the bandwidth, it also reduces the loop damping so as to slow the circuit response time. This may be undesirable. Bandwidth can be reduced by reducing the loop gain. This scheme will improve damping and permit faster operation under narrow-band operation. Note that the reduced impedance level at terminal 2 will require that a larger value of C_2 be used for a given filter cutoff frequency. If more than three 567's are to be used, the R_B, R_C network can be eliminated and the R_A resistors connected together. A capacitor between this junction and ground may be required to shunt high frequency components.

OUTPUT LATCHING

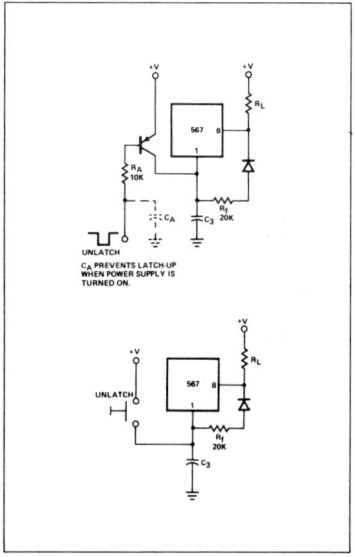

To latch the output on after a signal is received, it is necessary to provide a feedback resistor around the output stage (between pins 8 and 1). Pin 1 is pulled up to unlatch the output stage.

REDUCTION OF C_1 VALUE

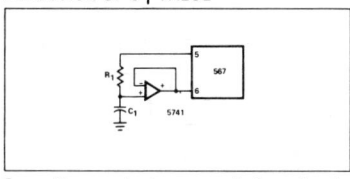

For precision, very low-frequency applications, where the value of C_1 becomes large, an overall cost savings may be achieved by inserting a voltage follower between the $R_1 C_1$ junction and pin 6, so as to allow a higher value of R_1 and a lower value of C_1 for a given frequency.

PROGRAMMING

To change the center frequency, the value of R_1 can be changed with a mechanical or solid state switch, or additional C_1 capacitors may be added by grounding them through saturating npn transistors.

Courtesy Signetics Corp.

193

567 – TONE DECODER PHASE LOCKED LOOP

TYPICAL APPLICATIONS

CARRIER-CURRENT REMOTE CONTROL OR INTERCOM

TOUCH-TONE® DECODER

DUAL-TONE DECODER

1. Resistor and capacitor values chosen for desired frequencies and bandwidth.
2. If C_3 is made large so as to delay turn-on of the top 567, decoding of sequential (f_1, f_2) tones is possible.

Component Values (Typical)

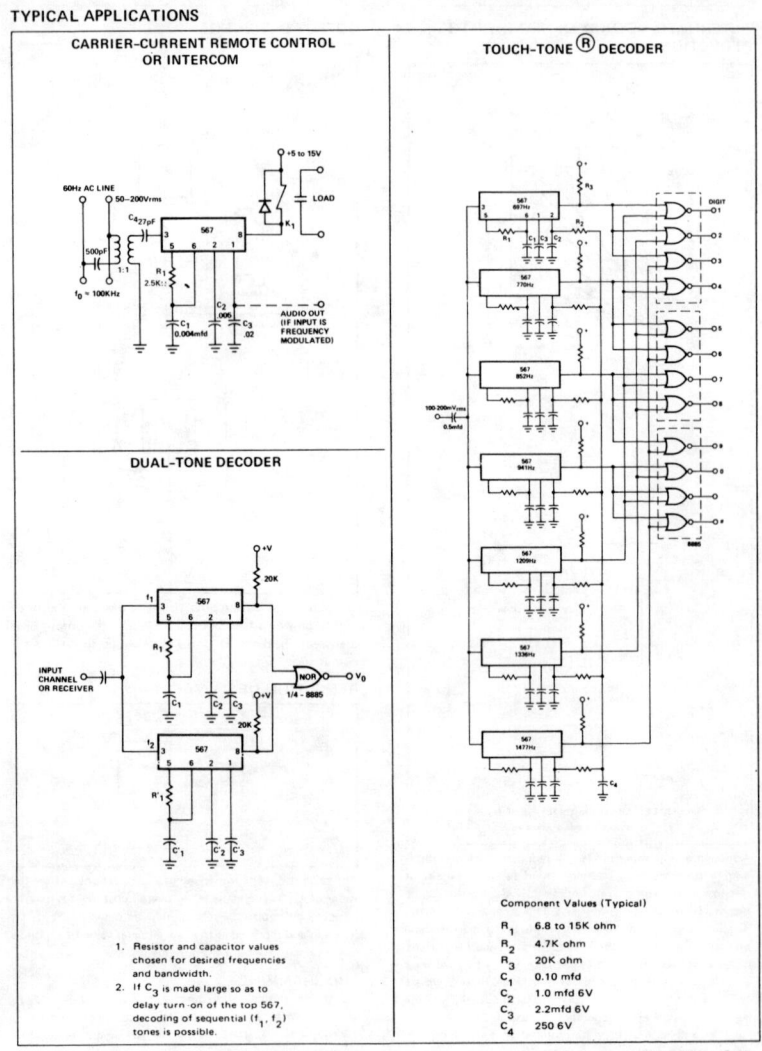

R_1	6.8 to 15K ohm
R_2	4.7K ohm
R_3	20K ohm
C_1	0.10 mfd
C_2	1.0 mfd 6V
C_3	2.2 mfd 6V
C_4	250 6V

Courtesy Signetics Corp.

567 – TONE DECODER PHASE LOCKED LOOP

TYPICAL APPLICATIONS (Cont'd.)

OSCILLATOR WITH QUADRATURE OUTPUT

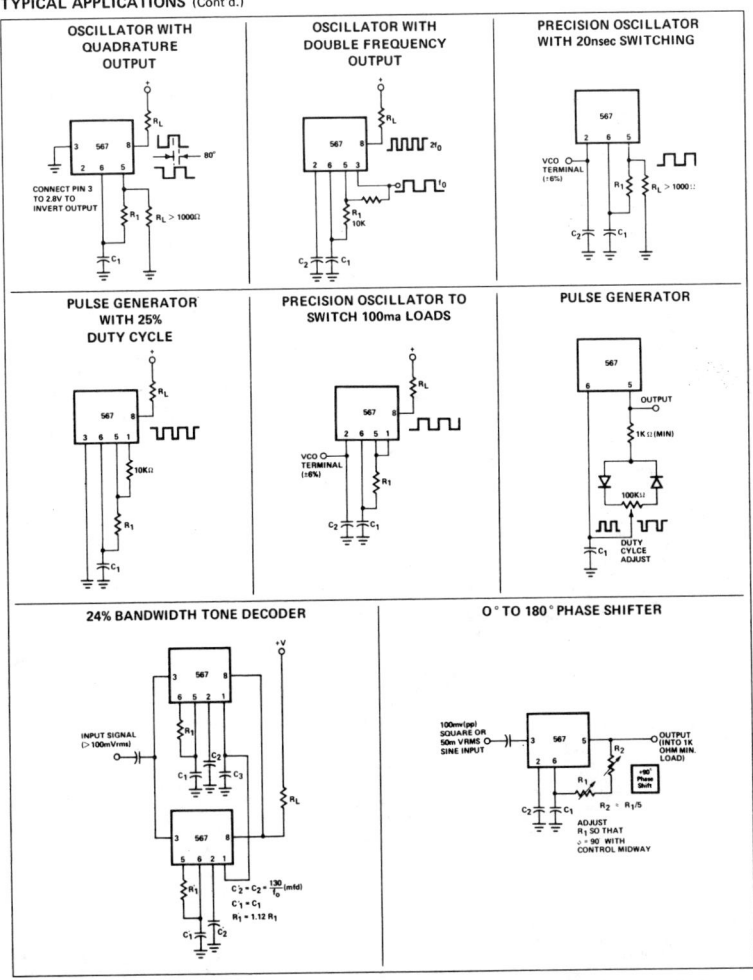

CONNECT PIN 3 TO 2.8V TO INVERT OUTPUT
$R_L > 1000\Omega$

OSCILLATOR WITH DOUBLE FREQUENCY OUTPUT

PRECISION OSCILLATOR WITH 20nsec SWITCHING

$R_L > 1000\Omega$

PULSE GENERATOR WITH 25% DUTY CYCLE

PRECISION OSCILLATOR TO SWITCH 100ma LOADS

VCO TERMINAL (±6%)

PULSE GENERATOR

OUTPUT
1KΩ (MIN)
100KΩ
DUTY CYCLE ADJUST

24% BANDWIDTH TONE DECODER

INPUT SIGNAL (>100mVrms)

$C'_2 = C_2 = \frac{130}{f_0}$ (mfd)
$C'_1 = C_1$
$R_1 = 1.12 R_1$

0° TO 180° PHASE SHIFTER

100mv(pp) SQUARE OR 50m VRMS SINE INPUT

OUTPUT (INTO 1K OHM MIN. LOAD)

$R_2 = R_1/5$

ADJUST R_1 SO THAT $\phi = 90°$ WITH CONTROL MIDWAY

Courtesy Signetics Corp.

195

CD4046A Types
COS/MOS Micropower Phase-Locked Loop

The RCA-CD4046A COS/MOS Micropower Phase-Locked Loop (PLL) consists of a low-power, linear voltage-controlled oscillator (VCO) and two different phase comparators having a common signal-input amplifier and a common comparator input. A 5.2-V zener diode is provided for supply regulation if necessary. The CD4046A is supplied in a 16-lead dual-in-line ceramic package (CD-4046AD), a 16-lead dual-in-line plastic package (CD4046AE), and a 16-lead flat pack (CD4046AK). It is also available in chip form (CD4046AH).

VCO Section

The VCO requires one external capacitor C1 and one or two external resistors (R1 or R1 and R2). Resistor R1 and capacitor C1 determine the frequency range of the VCO and resistor R2 enables the VCO to have a frequency offset if required. The high input impedance ($10^{12}\Omega$) of the VCO simplifies the design of low-pass filters by permitting the designer a wide choice of resistor-to-capacitor ratios. In order not to load the low-pass filter, a source-follower output of the VCO input voltage is provided at terminal 10 (DEMODULATED OUTPUT). If this terminal is used, a load resistor (R$_S$) of 10 kΩ or more should be connected from this terminal to V$_{SS}$. If unused this terminal should be left open. The VCO can be connected either directly or through frequency dividers to the comparator input of the phase comparators. A full COS/MOS logic swing is available at the output of the VCO and allows direct coupling to COS/MOS frequency dividers such as the RCA-CD4024, CD4018, CD4020, CD4022, CD4029, and CD4059. One or more CD4018 (Presettable Divide-by-N Counter) or CD4029 (Presettable Up/Down Counter), or CD4059A (Programmable Divide-by-"N" Counter), together with the CD4046A (Phase-Locked Loop) can be used to build a micropower low-frequency synthesizer. A logic 0 on the INHIBIT input "enables" the VCO and the source follower, while a logic 1 "turns off" both to minimize stand-by power consumption.

Phase Comparators

The phase-comparator signal input (terminal 14) can be direct-coupled provided the signal swing is within COS/MOS logic levels [logic "0" $\leq$ 30% (V$_{DD}$–V$_{SS}$), logic "1" $\geq$ 70% (V$_{DD}$–V$_{SS}$)]. For smaller swings the signal must be capacitively coupled to the self-biasing amplifier at the signal input.

Phase comparator I is an exclusive-OR network; it operates analogously to an over-driven balanced mixer. To maximize the lock range, the signal- and comparator-input frequencies must have a 50% duty cycle. With no signal or noise on the signal input, this phase comparator has an average output voltage equal to V$_{DD}$/2. The low-pass filter connected to the output of phase comparator I supplies the averaged voltage to the VCO input, and causes the VCO to oscillate at the center frequency (f$_o$).

The frequency range of input signals on which the PLL will lock if it was initially out of lock is defined as the frequency capture range (2f$_c$).

The frequency range of input signals on which the loop will stay locked if it was initially in lock is defined as the frequency lock range (2f$_L$). The capture range is $\leq$ the lock range.

With phase comparator I the range of frequencies over which the PLL can acquire lock (capture range) is dependent on the low-pass-filter characteristics, and can be made as large as the lock range. Phase-comparator I enables a PLL system to remain in lock in spite of high amounts of noise in the input signal.

One characteristic of this type of phase comparator is that it may lock onto input frequencies that are close to harmonics of the VCO center-frequency. A second characteristic is that the phase angle between the signal and the comparator input varies between 0° and 180°, and is 90° at the center frequency. Fig. 2 shows the typical, triangular, phase-to-output response characteristic

Features:
- Very low power consumption:
  70 μW (typ.) at VCO f$_o$ = 10 kHz, V$_{DD}$ = 5 V
- Operating frequency range up to 1.2 MHz (typ.) at V$_{DD}$ = 10 V
- Wide supply-voltage range: V$_{DD}$ – V$_{SS}$ = 5 to 15 V
- Low frequency drift: 0.06%/°C (typ.) at V$_{DD}$ = 10 V
- Choice of two phase comparators:
  1. Exclusive-OR network
  2. Edge-controlled memory network with phase-pulse output for lock indication
- High VCO linearity: 1% (typ.)
- VCO inhibit control for ON-OFF keying and ultra-low standby power consumption
- Source-follower output of VCO control input (Demod. output)
- Zener diode to assist supply regulation
- Quiescent current specified to 15 V
- Maximum input leakage current of 1 μA at 15 V (full package-temperature range)

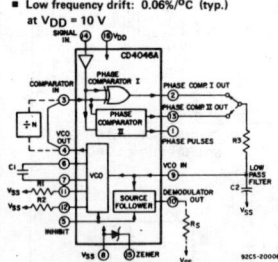

Applications:
- FM demodulator and modulator
- Frequency synthesis and multiplication
- Frequency discriminator
- Data synchronization
- Voltage-to-frequency conversion
- Tone decoding
- FSK – Modems
- Signal conditioning
- (See ICAN-6101) "RCA COS/MOS Phase-Locked Loop – A Versatile Building Block for Micropower Digital and Analog Applications"

Fig.1 — COS/MOS phase-locked loop block diagram.

MAXIMUM RATINGS, *Absolute-Maximum Values:*

STORAGE-TEMPERATURE RANGE (T$_{stg}$)	–65 to +150°C
OPERATING-TEMPERATURE RANGE (T$_A$)	
PACKAGE TYPES D, F, K, H	–55 to +125°C
PACKAGE TYPES E, Y	–40 to +85°C
DC SUPPLY-VOLTAGE RANGE, (V$_{DD}$)	
(Voltages referenced to V$_{SS}$ Terminal)	–0.5 to +15 V
POWER DISSIPATION PER PACKAGE (P$_D$):	
FOR T$_A$ = –40 to +60°C (PACKAGE TYPES E, Y)	500 mW
FOR T$_A$ = +60 to +85°C (PACKAGE TYPES E, Y)	Derate Linearly at 12 mW/°C to 200 mW
FOR T$_A$ = –55 to +100°C (PACKAGE TYPES D, F, K)	500 mW
FOR T$_A$ = +100 to +125°C (PACKAGE TYPES D, F, K)	Derate Linearly at 12 mW/°C to 200 mW
DEVICE DISSIPATION PER OUTPUT TRANSISTOR	
FOR T$_A$ = FULL PACKAGE-TEMPERATURE RANGE (ALL PACKAGE TYPES)	100 mW
INPUT VOLTAGE RANGE, ALL INPUTS	–0.5 to V$_{DD}$ +0.5 V
LEAD TEMPERATURE (DURING SOLDERING):	
At distance 1/16 ± 1/32 inch (1.59 ± 0.79 mm) from case for 10 s max.	+265°C

RECOMMENDED OPERATING CONDITIONS
For maximum reliability, nominal operating conditions should be selected so that operation is always within the following range:

CHARACTERISTIC	LIMITS		UNITS
	Min.	Max.	
Supply Voltage Range (For T$_A$ = Full Package Temperature Range	3	12	V

Courtesy RCA Solid State Div.

CD4046A Types

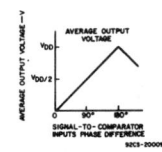

Fig.2 – Phase-comparator I characteristics at low-pass filter output.

of phase-comparator I. Typical waveforms for a COS/MOS phase-locked-loop employing phase comparator I in locked condition of f_o is shown in Fig. 3.

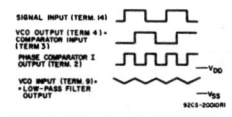

Fig.3 – Typical waveforms for COS/MOS phase-locked loop employing phase comparator I in locked condition of f_o.

Phase-comparator II is an edge-controlled digital memory network. It consists of four flip-flop stages, control gating, and a three-state output circuit comprising p- and n-type drivers having a common output node. When the p-MOS or n-MOS drivers are ON they pull the output up to V_{DD} or down to V_{SS}, respectively. This type of phase comparator acts only on the positive edges of the signal and comparator inputs. The duty cycles of the signal and comparator inputs are not important since positive transitions control the PLL system utilizing this type of comparator. If the signal-input frequency is higher than the comparator-input frequency, the p-type output driver is maintained ON most of the time, and both the n and p drivers OFF (3 state) the remainder of the time. If the signal-input frequency is lower than the comparator-input frequency, the n-type output driver is maintained ON most of the time, and both the n and p drivers OFF (3 state) the remainder of the time. If the signal- and comparator-input frequencies are the same, but the signal input lags the comparator input in phase, the n-type output driver is maintained ON for a time corresponding to the phase difference. If the signal- and comparator-input frequencies are the same, but the comparator input lags the signal in phase, the p-type output driver is maintained ON for a time corresponding to the phase difference. Subsequently, the capacitor voltage of the low-pass filter connected to this phase comparator is adjusted until the signal and comparator inputs are equal in both phase and frequency. At this stable point both p- and n-type output drivers remain OFF and thus the phase comparator output becomes an open circuit and holds the voltage on the capacitor of the low-pass filter constant.

ELECTRICAL CHARACTERISTICS at $T_A = 25°C$

Characteristic	Test Conditions		Limits All Package Types D,E,F,H,K,Y			Units
	V_O Volts	V_{DD} Volts	Min.	Typ.	Max.	

Phase Comparator Section

Characteristic	Test Conditions	V_O Volts	V_{DD} Volts	Min.	Typ.	Max.	Units
Operating Supply Voltage, $V_{DD} - V_{SS}$	VCO Operation			–	5	15	V
	Comparators only			–	3	15	
Total Quiescent Device Current, I_L: Term. 14 Open			5	–	25	55	µA
	Term. 15 open		10	–	200	410	
Term. 14 at V_{SS} or V_{DD}	Term. 5 at V_{DD}		5	–	5	15	
	Terms. 3 & 9 at V_{SS}		10	–	25	60	
			15	–	50	500	
Term. 14 (SIGNAL IN) Input Impedance, Z_{14}			5	1	2	–	MΩ
			10	0.2	0.4	–	
			15	–	0.2	–	
AC-Coupled Signal Input Voltage Sensitivity* (peak-to-peak)	See Fig.7		5	–	200	400	mV
			10	–	400	800	
			15	–	700	–	
DC-Coupled Signal Input and Comparator Input Voltage Sensitivity Low Level			5	1.5	2.25	–	V
			10	3	4.5	–	
			15	4.5	6.75	–	
High Level		V_O Volts	5	–	2.75	3.5	
			10	–	5.5	7	
			15	–	8.25	–	
Output Drive Current: n-Channel (Sink), I_{DN}	Phase Comparator I & II Term. 2 & 13	0.5	5	0.43	0.86	–	mA
		0.5	10	1.3	2.5	–	
	Phase Pulses	0.5	5	0.23	0.47	–	
		0.5	10	0.7	1.4	–	
p-Channel (Source), I_{DP}	Phase Comparator I & II Term. 2 & 13	4.5	5	–0.3	–0.6	–	
		9.5	10	–0.9	–1.8	–	
	Phase Pulses	4.5	5	–0.08	–0.16	–	
		9.5	10	–0.25	–0.5	–	
Input Leakage Current, I_{IL}, I_{IH} Max.	Any Input		15	–	±10⁻⁵	±1	µA

* For sine wave, the frequency must be greater than 1 kHz for Phase Comparator II.

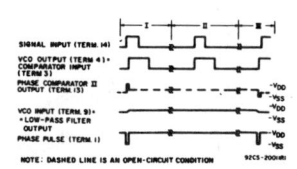

Fig.4 – Typical waveforms for COS/MOS phase-locked loop employing phase comparator II in locked condition.

Moreover the signal at the "phase pulses" output is a high level which can be used for indicating a locked condition. Thus, for phase comparator II, no phase difference exists between signal and comparator input over the full VCO frequency range. Moreover, the power dissipation due to the low-pass filter is reduced when this type of phase comparator is used because both the p- and n-type output drivers are OFF for most of the signal input cycle. It should be noted that the PLL lock range for this type of phase comparator is equal to the capture range, independent of the low-pass filter. With no signal present at the signal input, the VCO is adjusted to its lowest frequency for phase comparator II. Fig. 4 shows typical waveforms for a COS/MOS PLL employing phase comparator II in a locked condition.

Courtesy RCA Solid State Div.

CD4046A Types

DESIGN INFORMATION

This information is a guide for approximating the values of external components for the CD4046A in a Phase-Locked-Loop system. The selected external components must be within the following ranges:

$10\ k\Omega \leq R1, R2, R_S \leq 1\ M\Omega$
$C1 \geq 100\ pF$ at $V_{DD} \geq 5\ V$;
$C1 \geq 50\ pF$ at $V_{DD} \geq 10\ V$

In addition to the given design information refer to Fig.5 for R1, R2, and C1 component selections.

Characteristics	Phase Comparator Used	Design Information	
		VCO WITHOUT OFFSET $R2 = \infty$	VCO WITH OFFSET
VCO Frequency	1	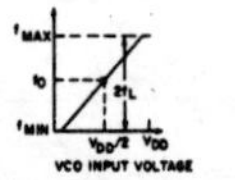	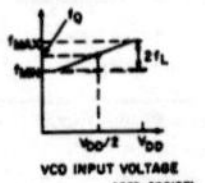
For No Signal Input	2	Same as for No.1	
	1	VCO will adjust to center frequency, f_o	
	2	VCO will adjust to lowest operating frequency, f_{min}	
Frequency Lock Range, $2 f_L$	1	$2 f_L$ = full VCO frequency range	
		$2 f_L = f_{max} - f_{min}$	
	2	Same as for No.1	
Frequency Capture Range, $2 f_C$	1	$\tau_1 = R3C2$ (1), (2) $2 f_C \approx \frac{1}{\pi}\sqrt{\frac{2\pi f_L}{\tau_1}}$	
Loop Filter Component Selection	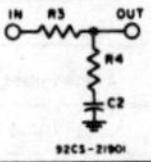	For $2 f_C$, see Ref. (2)	
	2	$f_C = f_L$	
Phase Angle Between Signal and Comparator	1	$90°$ at center frequency (f_o) approximating $0°$ and $180°$ at ends of lock-range ($2 f_L$)	
	2	Always $0°$ in lock	

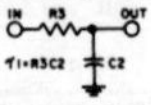

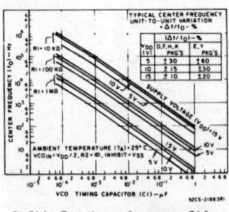

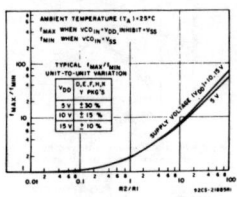

Fig.5(c) – Typical f_{max}/f_{min} vs R2/R1.

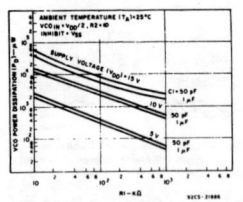

Fig.6(a) – Typical VCO power dissipation at center frequency vs R1.

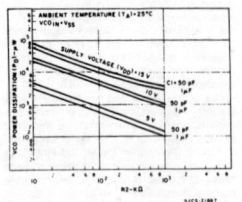

Fig.6(b) – Typical VCO power dissipation at f_{min} vs R2.

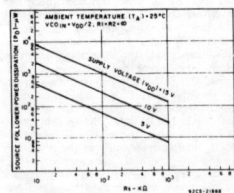

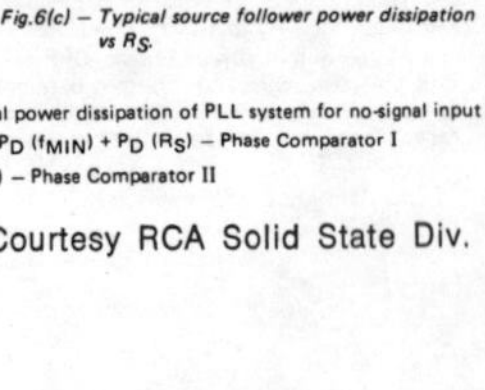

Fig.6(c) – Typical source follower power dissipation vs R_S.

Fig.5(a) – Typical center frequency vs C1 for $R1 = 10\ k\Omega$, and $1\ M\Omega$ and $f_o \sim 1/R1\ C1$.

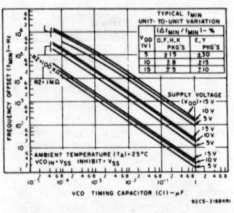

Fig.5(b) – Typical frequency offset vs C1 for $R2 = 10\ k\Omega$, $100\ k\Omega$, and $1\ M\Omega$.

NOTE: Lower frequency values are obtainable if larger values of C1 than shown in Figs. 5(a) and 5(b) are used.

NOTE: To obtain approximate total power dissipation of PLL system for no-signal input
P_D (Total) = P_D (f_o) + P_D (f_{MIN}) + P_D (R_S) – Phase Comparator I
P_D (Total) = P_D (f_{MIN}) – Phase Comparator II

Courtesy RCA Solid State Div.

CD4046A Types

DESIGN INFORMATION (Cont'd):

Characteristics	Phase Comparator Used	Design Information
Locks On Harmonic of Center Frequency	1	Yes
	2	No
Signal Input Noise Rejection	1	High
	2	Low
VCO Component Selection	1	**VCO WITHOUT OFFSET** $R_2 = \infty$ – Given: f_o – Use f_o with Fig.5a to determine R1 and C1	**VCO WITH OFFSET** – Given: f_o and f_L – Calculate f_{min} from the equation $f_{min} = f_o - f_L$ – Use f_{min} with Fig.5b to determine R2 and C1 – Calculate $\dfrac{f_{max}}{f_{min}}$ from the equation $\dfrac{f_{max}}{f_{min}} = \dfrac{f_o + f_L}{f_o - f_L}$ – Use $\dfrac{f_{max}}{f_{min}}$ with Fig.5c to determine ratio R2/R1 to obtain R1
	2	– Given: f_{max} – Calculate f_o from the equation $f_o = \dfrac{f_{max}}{2}$ –Use f_o with Fig.5a to determine R1 and C1	– Given: f_{min} & f_{max} – Use f_{min} with Fig.5b to determine R2 and C1 – Calculate $\dfrac{f_{max}}{f_{min}}$ – Use $\dfrac{f_{max}}{f_{min}}$ with Fig.5c to determine ratio R2/R1 to obtain R1

For further information, see
(1) F. Gardner, "Phase-Lock Techniques" John Wiley and Sons, New York, 1966
(2) G. S. Moschytz, "Miniaturized RC Filters Using Phase-Locked Loop", BSTJ, May, 1965.

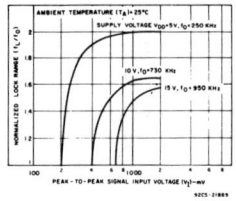

Fig.7 – Typical lock range vs signal input amplitude.

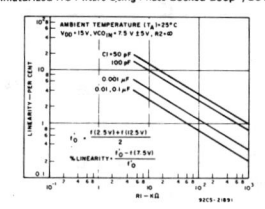

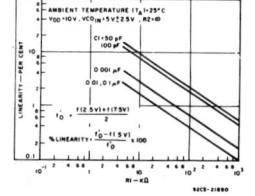

Fig.8(a) and (b) – Typical VCO linearity vs R1 and C1.

Courtesy RCA Solid State Div.

CD4046A Types

ELECTRICAL CHARACTERISTICS at $T_A = 25°C$

Characteristic	Test Conditions			Limits All Package Types D,E,F,H,K,Y			Units
		V_O Volts	V_{DD} Volts	Min.	Typ.	Max.	
VCO Section							
Operating Supply Voltage $V_{DD} - V_{SS}$	As fixed oscillator only			3	–	15	V
	Phase-lock-loop operation			5	–	15	
Operating Power Dissipation, P_D	$f_o = 10$ kHz, $R_1 = 1$ MΩ, $R_2 = \infty$, $VCO_{IN} = \frac{V_{DD}}{2}$		5	–	70	–	μW
			10	–	600	–	
			15	–	2400	–	
Maximum Operating Frequency, f_{max}	$R_1 = 10$ kΩ, $C_1 = 100$ pF, $R_2 = \infty$, $C_1 = 50$ pF, $VCO_{IN} = V_{DD}$		5	0.25	0.5	–	MHz
			10	0.6	1.2	–	
			15	–	1.5	–	
Center Frequency (f_o) and Frequency Range, $f_{max} - f_{min}$	Programmable with external components R1, R2, and C1 See Design Information						
Linearity	$VCO_{IN} = 2.5$ V ± 0.3 V, R1 > 10 kΩ		5	–	1	–	%
	= 5 V ± 2.5 V, R1 > 400 kΩ		10	–	1	–	
	= 7.5 V ± 5 V, R1 = 1 MΩ		15	–	1	–	
Temperature-Frequency Stability●: No Frequency Offset $f_{MIN} = 0$	%/°C $\alpha \frac{1}{f \cdot V_{DD}}$ $R_2 = \infty$		5	–	0.12–0.24	–	%/°C
			10	–	0.04–0.08	–	
			15	–	0.015–0.03	–	
Frequency Offset $f_{MIN} \ne 0$	%/°C $\alpha \frac{1}{f \cdot V_{DD}}$		5	–	0.06–0.12	–	
			10	–	0.05–0.1	–	
			15	–	0.03–0.06	–	
Input Resistance of VCO_{IN} (Term 9), R_I			5,10,15	–	10^{12}	–	Ω
VCO Output Voltage (Term 4) Low Level, V_{OL}	Driving COS/MOS-Type Load (e.g. Term 3 Phase Comparator Input)		5,10,15	–	–	0.01	V
High Level, V_{OH}			5	4.99	–	–	
			10	9.99	–	–	
			15	14.99	–	–	
VCO Output Duty Cycle			5,10,15	–	50	–	%
VCO Output Transition Times, t_{THL}, t_{TLH}		V_O Volts	5	–	75	150	ns
			10	–	50	100	
			15	–	40	–	
VCO Output Drive Current: n-Channel (Sink), I_{DN}		0.5	5	0.43	0.86	–	mA
		0.5	10	1.3	2.6	–	
p-Channel (Source), I_{DP}		4.5	5	–0.3	–0.6	–	
		9.5	10	–0.9	–1.8	–	
Source-Follower Output (Demodulated Output): Offset Voltage ($VCO_{IN} - V_{DEM}$)	$R_S > 10$ kΩ		5,10	–	1.5	2.2	V
			15	–	1.5	–	
Linearity	$R_S > 50$ kΩ, $VCO_{IN} = 2.5 \pm 0.3$ V		5	–	0.1	–	%
	= 5 ± 2.5 V		10	–	0.6	–	
	= 7.5 ± 5 V		15	–	0.8	–	
Zener Diode Voltage (V_Z): CD4046AD,AF,AK	$I_Z = 50$ μA			4.7	5.2	5.7	V
CD4046AE,AY				4.5	5.2	6.1	
Zener Dynamic Resistance, R_Z	$I_Z = 1$ mA			–	100	–	Ω

● Positive coefficient.

Courtesy RCA Solid State Div.

MC1648
MC1648M

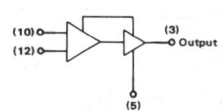

The MC1648 is an emitter-coupled oscillator, constructed on a single monolithic silicon chip. Output levels are compatible with MECL III logic levels. The oscillator requires an external parallel tank circuit consisting of the inductor (L) and capacitor (C).

A varactor diode may be incorporated into the tank circuit to provide a voltage variable input for the oscillator (VCO). The MC1648 was designed for use in the Motorola Phase-Locked Loop shown in Figure 9. This device may also be used in many other applications requiring a fixed or variable frequency clock source of high spectral purity (See Figure 2).

The MC1648 may be operated from a +5.0 Vdc supply or a -5.2 Vdc supply, depending upon system requirements.

Numbers in parenthesis denote pin number for F package (Case 607) L package (Case 632), and P package (Case 646).

Input Capacitance = 6 pF typ
Maximum Series Resistance for L (External Inductance) = 50 Ω typ
Power Dissipation = 150 mW typ/pkg (+5.0 Vdc Supply)
Maximum Output Frequency = 225 MHz typ

SUPPLY VOLTAGE	GND PINS	SUPPLY PINS
+5.0 Vdc	7, 8	1, 14
-5.2 Vdc	1, 14	7, 8

FIGURE 1 – CIRCUIT SCHEMATIC

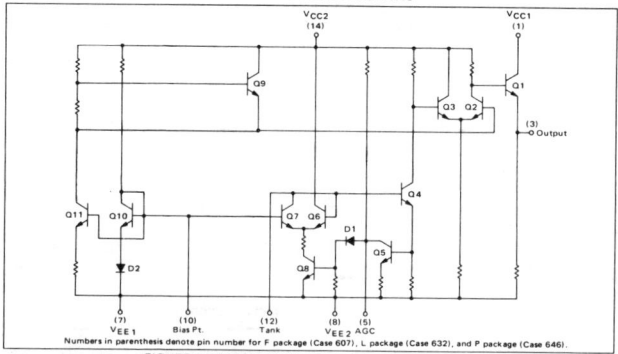

FIGURE 2 – SPECTRAL PURITY OF SIGNAL AT OUTPUT

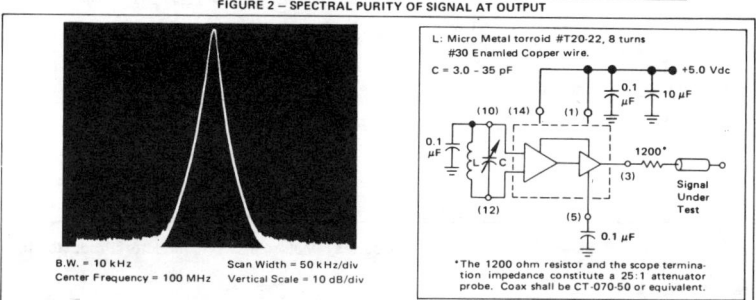

B.W. = 10 kHz
Center Frequency = 100 MHz
Scan Width = 50 kHz/div
Vertical Scale = 10 dB/div

L: Micro Metal torroid #T20-22, 8 turns #30 Enamled Copper wire.
C = 3.0 - 35 pF

*The 1200 ohm resistor and the scope termination impedance constitute a 25:1 attenuator probe. Coax shall be CT-070-50 or equivalent.

Courtesy Motorola Semiconductor Products, Inc.

201

ELECTRICAL CHARACTERISTICS

Supply Voltage = +5.0 volts

			TEST VOLTAGE/CURRENT VALUES			
			(Volts)			mAdc
		@ Test Temperature	$V_{IH\,max}$	$V_{IL\,min}$	V_{CC}	I_L
		$-30°C$	+2.00	+1.50	5.0	-5.0
		$+25°C$	+1.85	+1.35	5.0	-5.0
		$+85°C$	+1.70	+1.20	5.0	-5.0

Characteristic	Symbol	Pin Under Test	MC1648 Test Limits						Unit	TEST VOLTAGE/CURRENT APPLIED TO PINS LISTED BELOW:				V_{EE} (Gnd)
			-30°C		+25°C		+85°C			$V_{IH\,max}$	$V_{IL\,min}$	V_{CC}	I_L	
			Min	Max	Min	Max	Min	Max						
Power Supply Drain Current	I_E	8				40			mAdc			1, 14		7, 8
Logic "1" Output Voltage	V_{OH}	3	3.955	4.185	4.04	4.25	4.11	4.36	Vdc		12	1, 14	3	7, 8
Logic "0" Output Voltage	V_{OL}	3	3.16	3.40	3.20	3.43	3.22	3.475	Vdc	12		1, 14	3	7, 8
Bias Voltage	V_{Bias}*	10	1.60	1.90	1.45	1.75	1.30	1.60	Vdc		12	1, 14		7, 8
			Min	Typ	Max	Min	Typ	Max	Min	Typ	Max			
Peak-to-Peak Tank Voltage	V_{p-p}	12					400					mV	See Figure 3		1, 14	3	7, 8
Output Duty Cycle	V_{DC}	3					50					%	See Figure 3		1, 14	3	7, 8
Oscillation Frequency	f_{max}	—	—	225	—	200	225	—	—	225	—	MHz	See Figure 3	—	1, 14	3	7, *

*This measurement guarantees the dc potential at the bias point for purposes of incorporating a varactor tuning diode at this point.

V_{p-p} output is typically 500 mV @ 225 MHz.

ELECTRICAL CHARACTERISTICS

Supply Voltage = -5.2 volts

			TEST VOLTAGE/CURRENT VALUES			
			(Volts)			mAdc
		@ Test Temperature	$V_{IH\,max}$	$V_{IL\,min}$	V_{EE}	I_L
		$-30°C$	-3.20	-3.70	-5.2	-5.0
		$+25°C$	-3.35	-3.85	-5.2	-5.0
		$+85°C$	-3.500	-4.000	-5.2	-5.0

Characteristic	Symbol	Pin Under Test	MC1648 Test Limits						Unit	TEST VOLTAGE/CURRENT APPLIED TO PINS LISTED BELOW:				V_{CC} (Gnd)
			-30°C		+25°C		+85°C			$V_{IH\,max}$	$V_{IL\,min}$	V_{EE}	I_L	
			Min	Max	Min	Max	Min	Max						
Power Supply Drain Current	I_E	8				41			mAdc			7, 8		1, 14
Logic "1" Output Voltage	V_{OH}	3	-1.045	-0.815	-0.960	-0.750	-0.890	-0.640	Vdc		12	7, 8	3	1, 14
Logic "0" Output Voltage	V_{OL}	3	-1.890	-1.650	-1.850	-1.620	-1.830	-1.575	Vdc	12		7, 8	3	1, 14
Bias Voltage	V_{Bias}*	10	-3.60	-3.30	-3.75	-3.45	-3.90	-3.60	Vdc		12	7, 8		1, 14
			Min	Typ	Max	Min	Typ	Max	Min	Typ	Max			
Peak-to-Peak Tank Voltage	V_{p-p}	12					400					mV	See Figure 3		7, 8	3	1, 14
Output Duty Cycle	V_{DC}	3					50					%	See Figure 3		7, 8	3	1, 14
Oscillation Frequency	f_{max}	—	—	225	—	200	225	—	—	225	—	MHz	See Figure 3	—	7, 8	3	1, 14

*This measurement guarantees the dc potential at the bias point for purposes of incorporating a varactor tuning diode at this point.

V_{p-p} output is typically 500 mV @ 225 MHz.

FIGURE 3 — TEST CIRCUIT AND WAVEFORMS

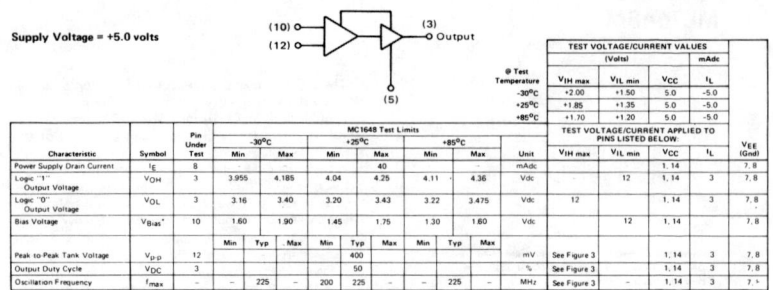

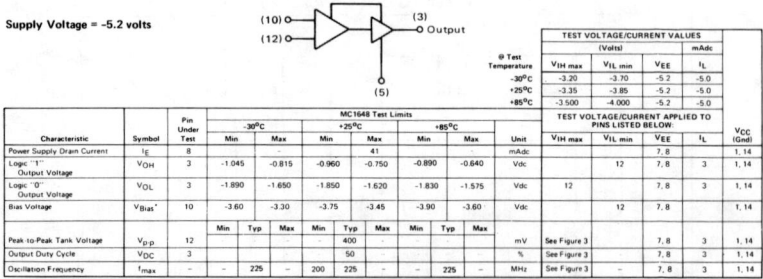

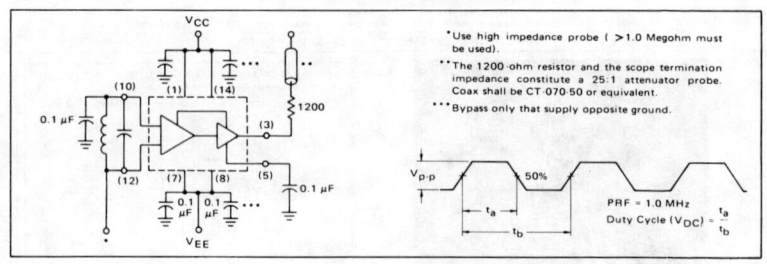

* Use high impedance probe (>1.0 Megohm must be used).
** The 1200-ohm resistor and the scope termination impedance constitute a 25:1 attenuator probe. Coax shall be CT-070-50 or equivalent.
*** Bypass only that supply opposite ground.

PRF = 1.0 MHz
Duty Cycle $(V_{DC}) = \dfrac{t_a}{t_b}$

Courtesy Motorola Semiconductor Products, Inc.

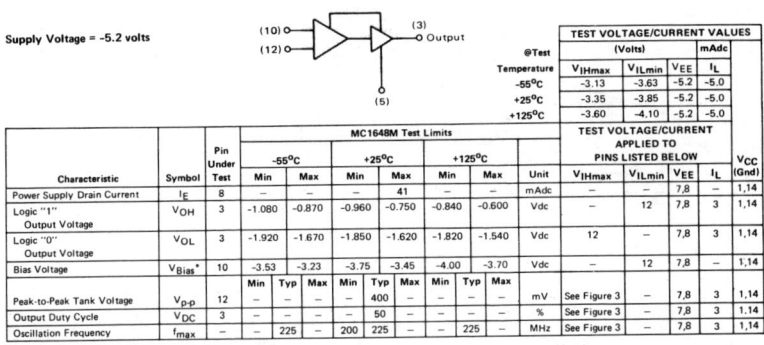

ELECTRICAL CHARACTERISTICS

Supply Voltage = +5.0 volts

Test circuit pins: (10), (12) inputs; (5) bias; (3) Output

@Test Temperature	TEST VOLTAGE/CURRENT VALUES (Volts)			mAdc
	V_{IHmax}	V_{ILmin}	V_{CC}	I_L
-55°C	+2.07	+1.57	5.0	-5.0
+25°C	+1.85	+1.35	5.0	-5.0
+125°C	+1.60	+1.10	5.0	-5.0

Characteristic	Symbol	Pin Under Test	-55°C Min	-55°C Max	+25°C Min	+25°C Max	+125°C Min	+125°C Max	Unit	TEST VOLTAGE/CURRENT APPLIED TO PINS LISTED BELOW V_{IHmax}	V_{ILmin}	V_{CC}	I_L	V_{EE} (Gnd)
Power Supply Drain Current	I_E	8	–	–	–	40	–	–	mAdc	–	–	1,14	–	7,8
Logic "1" Output Voltage	V_{OH}	3	3.92	4.13	4.04	4.25	4.16	4.40	Vdc	–	12	1,14	3	7,8
Logic "0" Output Voltage	V_{OL}	3	3.13	3.38	3.20	3.43	3.23	3.51	Vdc	12	–	1,14	3	7,8
Bias Voltage	V_{Bias}*	10	1.67	1.97	1.45	1.75	1.20	1.50	Vdc	–	12	1,14	–	7,8
			Min	Typ	Max	Min	Typ	Max	Min	Typ	Max			
Peak-to-Peak Tank Voltage	V_{p-p}	12	–	–	–	–	400	–	–	–	–	mV	See Figure 3	–	1,14	3	7,8
Output Duty Cycle	V_{DC}	3	–	–	–	–	50	–	–	–	–	%	See Figure 3	–	1,14	3	7,8
Oscillation Frequency	f_{max}	–	–	225	–	200	225	–	–	225	–	MHz	See Figure 3	–	1,14	3	7,8

*This measurement guarantees the dc potential at the bias point for purposes of incorporating a varactor tuning diode at this point.
V_{p-p} output is typically 500 mV @ 225 MHz.

ELECTRICAL CHARACTERISTICS

Supply Voltage = -5.2 volts

@Test Temperature	TEST VOLTAGE/CURRENT VALUES (Volts)			mAdc
	V_{IHmax}	V_{ILmin}	V_{EE}	I_L
-55°C	-3.13	-3.63	-5.2	-5.0
+25°C	-3.35	-3.85	-5.2	-5.0
+125°C	-3.60	-4.10	-5.2	-5.0

Characteristic	Symbol	Pin Under Test	-55°C Min	-55°C Max	+25°C Min	+25°C Max	+125°C Min	+125°C Max	Unit	TEST VOLTAGE/CURRENT APPLIED TO PINS LISTED BELOW V_{IHmax}	V_{ILmin}	V_{EE}	I_L	V_{CC} (Gnd)
Power Supply Drain Current	I_E	8	–	–	–	41	–	–	mAdc	–	–	7,8	–	1,14
Logic "1" Output Voltage	V_{OH}	3	-1.080	-0.870	-0.960	-0.750	-0.840	-0.600	Vdc	–	12	7,8	3	1,14
Logic "0" Output Voltage	V_{OL}	3	-1.920	-1.670	-1.850	-1.620	-1.820	-1.540	Vdc	12	–	7,8	3	1,14
Bias Voltage	V_{Bias}*	10	-3.53	-3.23	-3.75	-3.45	-4.00	-3.70	Vdc	–	12	7,8	–	1,14
			Min	Typ	Max	Min	Typ	Max	Min	Typ	Max			
Peak-to-Peak Tank Voltage	V_{p-p}	12	–	–	–	–	400	–	–	–	–	mV	See Figure 3	–	7,8	3	1,14
Output Duty Cycle	V_{DC}	3	–	–	–	–	50	–	–	–	–	%	See Figure 3	–	7,8	3	1,14
Oscillation Frequency	f_{max}	–	–	225	–	200	225	–	–	225	–	MHz	See Figure 3	–	7,8	3	1,14

*This measurement guarantees the dc potential at the bias point for purposes of incorporating a varactor tuning diode at this point.
V_{p-p} output is typically 500 mV @ 225 MHz.

Courtesy Motorola Semiconductor Products, Inc.

OPERATING CHARACTERISTICS

Figure 1 illustrates the circuit schematic for the MC1648. The oscillator incorporates positive feedback by coupling the base of transistor Q6 to the collector of Q7. An automatic gain control (AGC) is incorporated to limit the current through the emitter-coupled pair of transistors (Q6 and Q7) and allow optimum frequency response of the oscillator.

In order to maintain the high Q of the oscillator, and provide high spectral purity at the output, transistor Q4 is used to translate the oscillator signal to the output differential pair Q2 and Q3. Q2 and Q3, in conjunction with output transistor Q1, provide a highly buffered output which produces a square wave. Transistors Q9 thru Q11 provide the bias drive for the oscillator and output buffer. Figure 2 indicates the high spectral purity of the oscillator output (pin 3).

When operating the oscillator in the voltage controlled mode (Figure 4), it should be noted that the cathode of the varactor diode (D) should be biased at least 2 V_{BE} above V_{EE} ($\approx$ 1.4 V for positive supply operation).

FIGURE 4 — THE MC1648 OPERATING IN THE VOLTAGE CONTROLLED MODE

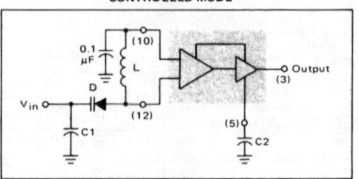

When the MC1648 is used with a constant dc voltage to the varactor diode, the output frequency will vary slightly because of internal noise. This variation is plotted versus operating frequency in Figure 5.

FIGURE 5 — NOISE DEVIATION TEST CIRCUIT AND WAVEFORM

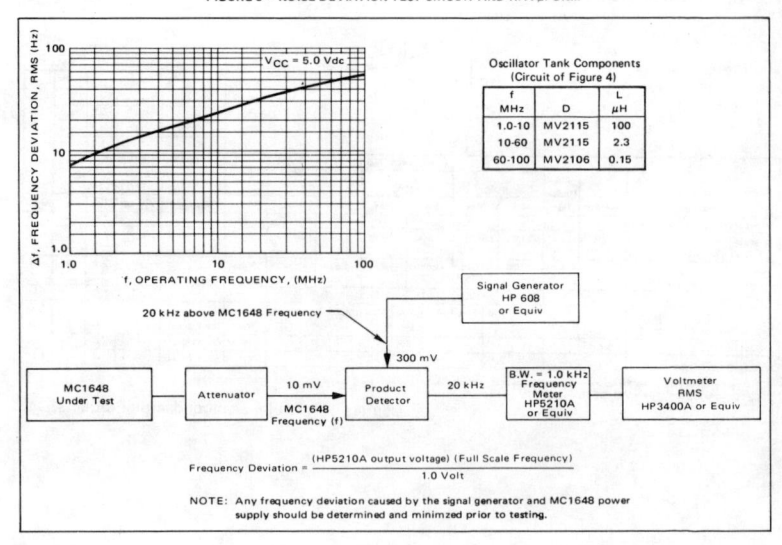

Oscillator Tank Components
(Circuit of Figure 4)

f MHz	D	L µH
1.0-10	MV2115	100
10-60	MV2115	2.3
60-100	MV2106	0.15

Frequency Deviation = (HP5210A output voltage) (Full Scale Frequency) / 1.0 Volt

NOTE: Any frequency deviation caused by the signal generator and MC1648 power supply should be determined and minimzed prior to testing.

Courtesy Motorola Semiconductor Products, Inc.

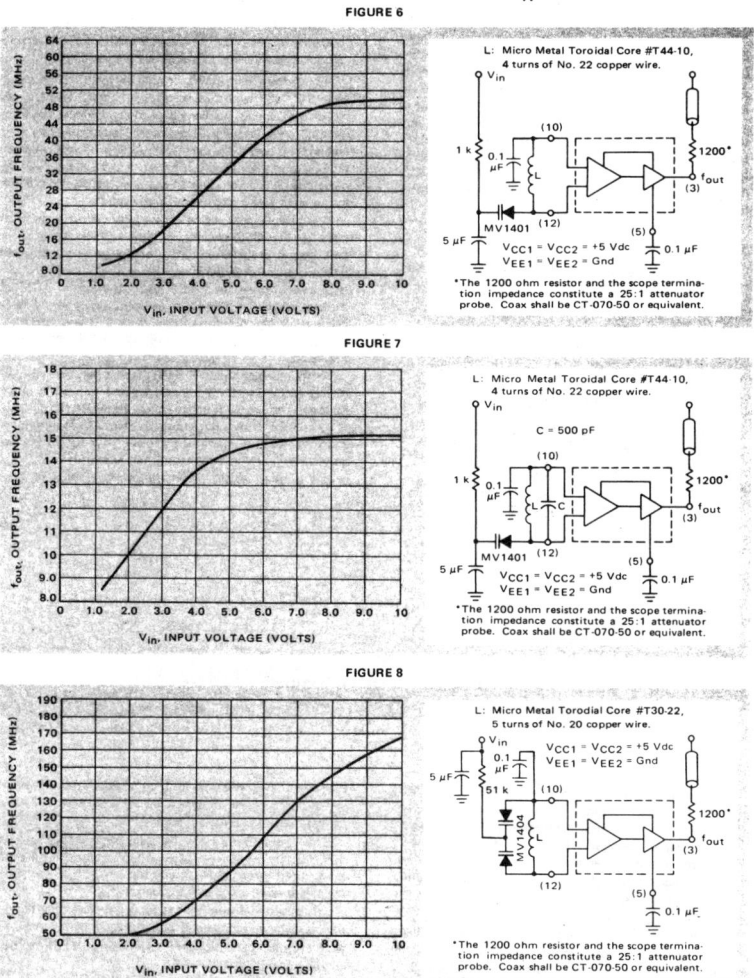

TRANSFER CHARACTERISTICS IN THE VOLTAGE CONTROLLED MODE
USING EXTERNAL VARACTOR DIODE AND COIL. $T_A = 25°C$

FIGURE 6

L: Micro Metal Toroidal Core #T44-10, 4 turns of No. 22 copper wire.

$V_{CC1} = V_{CC2} = +5$ Vdc
$V_{EE1} = V_{EE2} =$ Gnd

*The 1200 ohm resistor and the scope termination impedance constitute a 25:1 attenuator probe. Coax shall be CT-070-50 or equivalent.

FIGURE 7

L: Micro Metal Toroidal Core #T44-10, 4 turns of No. 22 copper wire.
C = 500 pF

$V_{CC1} = V_{CC2} = +5$ Vdc
$V_{EE1} = V_{EE2} =$ Gnd

*The 1200 ohm resistor and the scope termination impedance constitute a 25:1 attenuator probe. Coax shall be CT-070-50 or equivalent.

FIGURE 8

L: Micro Metal Torodial Core #T30-22, 5 turns of No. 20 copper wire.

$V_{CC1} = V_{CC2} = +5$ Vdc
$V_{EE1} = V_{EE2} =$ Gnd

*The 1200 ohm resistor and the scope termination impedance constitute a 25:1 attenuator probe. Coax shall be CT-070-50 or equivalent.

Courtesy Motorola Semiconductor Products, Inc.

205

Typical transfer characteristics for the oscillator in the voltage controlled mode are shown in Figures 6, 7 and 8. Figures 6 and 8 show transfer characteristics employing only the capacitance of the varactor diode (pluse the input capacitance of the oscillator, 6 pF typical). Figure 7 illustrates the oscillator operating in a voltage controlled mode with the output frequency range limited. This is achieved by adding a capacitor in parallel with the tank circuit as shown. The 1 kΩ resistor in Figures 6 and 7 is used to protect the varactor diode during testing. It is not necessary as long as the dc input voltage does not cause the diode to become forward biased. The larger-valued resistor (51 kΩ) in Figure 8 is required to provide isolation for the high-impedance junctions of the two varactor diodes.

The tuning range of the oscillator in the voltage controlled mode may be calculated as:

$$\frac{f_{max}}{f_{min}} = \frac{\sqrt{C_D \text{ (max)} + C_S}}{\sqrt{C_D \text{ (min)} + C_S}}$$

where $f_{min} = \dfrac{1}{2\pi \sqrt{L \left(C_D \text{ (max)} + C_S\right)}}$

C_S = shunt capacitance (input plus external capacitance).

C_D = varactor capacitance as a function of bias voltage.

Good RF and low-frequency bypassing is necessary on the power supply pins (see Figure 2).

Capacitors (C1 and C2 of Figure 4) should be used to bypass the AGC point and the VCO input (varactor diode), guaranteeing only dc levels at these points.

For output frequency operation between 1 MHz and 50 MHz a 0.1 µF capacitor is sufficient for C1 and C2. At higher frequencies, smaller values of capacitance should be used; at lower frequencies, larger values of capacitance. At higher frequencies the value of bypass capacitors depends directly upon the physical layout of the system. All bypassing should be as close to the package pins as possible to minimize unwanted lead inductance.

The peak-to-peak swing of the tank circuit is set internally by the AGC circuitry. Since voltage swing of the tank circuit provides the drive for the output buffer, the AGC potential directly affects the output waveform. If it is desired to have a sine wave at the output of the MC1648, a series resistor is tied from the AGC point to the most negative power potential (ground if +5.0 volt supply is used, -5.2 volts if a negative supply is used) as shown in Figure 10.

At frequencies above 100 MHz typ, it may be desirable to increase the tank circuit peak-to-peak voltage in order to shape the signal at the output of the MC1648. This is accomplished by tying a series resistor (1 kΩ minimum) from the AGC to the most positive power potential (+5.0 volts if a +5.0 volt supply is used, ground if a -5.2 volt supply is used). Figure 11 illustrates this principle.

APPLICATIONS INFORMATION

The phase locked loop shown in Figure 9 illustrates the use of the MC1648 as a voltage controlled oscillator. The figure illustrates a frequency synthesizer useful in tuners for FM broadcast, general aviation, maritime and land-mobile communications, amateur and CB receivers. The system operates from a single +5.0 Vdc supply, and requires no internal translation, since all components are compatible. Frequency generation of this type offers the advantages of single crystal operation, simple channel selection, and elimination of special circuitry to prevent harmonic lock-up. Additional features include dc digital switching (preferable over RF switching with a multiple crystal system), and a broad range of tuning (up to 150 MHz, the range being set by the varactor diode).

The output frequency of the synthesizer loop is determined by the reference frequency and the number programmed at the programmable counter; $f_{out} = Nf_{ref}$. The channel spacing is equal to frequency (f_{ref}).

For additional information on applications and designs for phase locked-loops and digital frequency synthesizers, see Motorola Application Notes AN-532A, AN-535, AN-553, AN-564 or AN-594.

Courtesy Motorola Semiconductor Products, Inc.

206

FIGURE 9 — TYPICAL FREQUENCY SYNTHESIZER APPLICATION

$f_{out} = N f_{ref}$ where $N = N_p \bullet P + A$

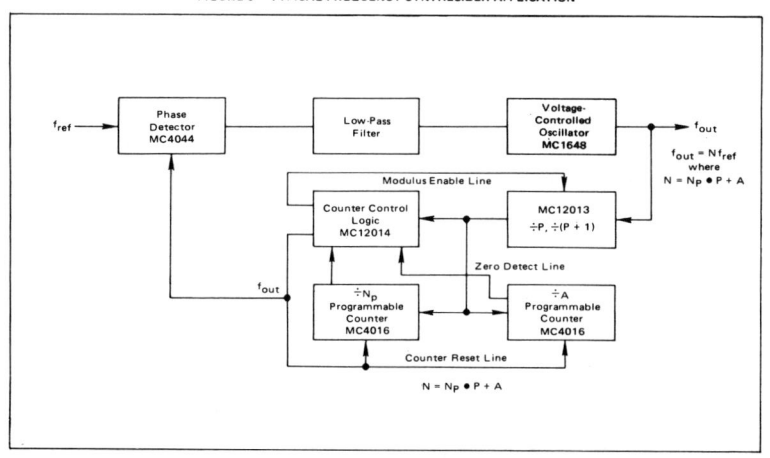

Figure 10 shows the MC1648 in the variable frequency mode operating from a +5.0 Vdc supply. To obtain a sine wave at the output, a resistor is added from the AGC circuit (pin 5) to V_{EE}.

Figure 11 shows the MC1648 in the variable frequency mode operating from a +5.0 Vdc supply. To extend the useful frequency range of the device a resistor is added to the AGC circuit at pin 5 (1 kohm minimum).

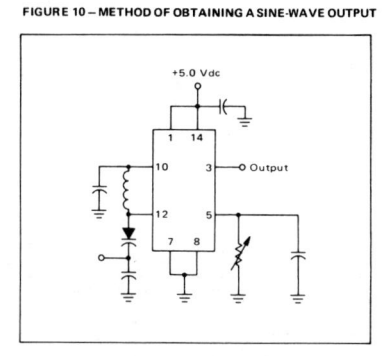

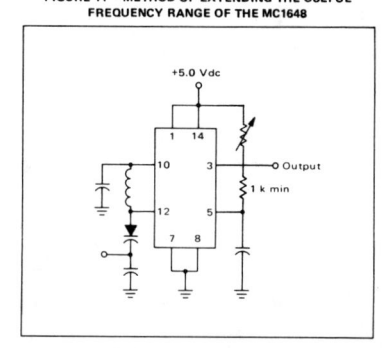

Figure 12 shows the MC1648 operating from +5.0 Vdc and +9.0 Vdc power supplies. This permits a higher voltage swing and higher output power than is possible from the MECL output (pin 3). Plots of output power versus total collector load resistance at pin 1 are given in Figures 13 and 14 for 100 MHz and 10 MHz operation. The total collector load includes R in parallel with Rp of L1 and C1 at resonance. The optimum value for R at 100 MHz is approximately 850 ohms.

FIGURE 10 — METHOD OF OBTAINING A SINE-WAVE OUTPUT

FIGURE 11 — METHOD OF EXTENDING THE USEFUL FREQUENCY RANGE OF THE MC1648

Courtesy Motorola Semiconductor Products, Inc.

FIGURE 12 — CIRCUIT SCHEMATIC USED FOR COLLECTOR OUTPUT OPERATION

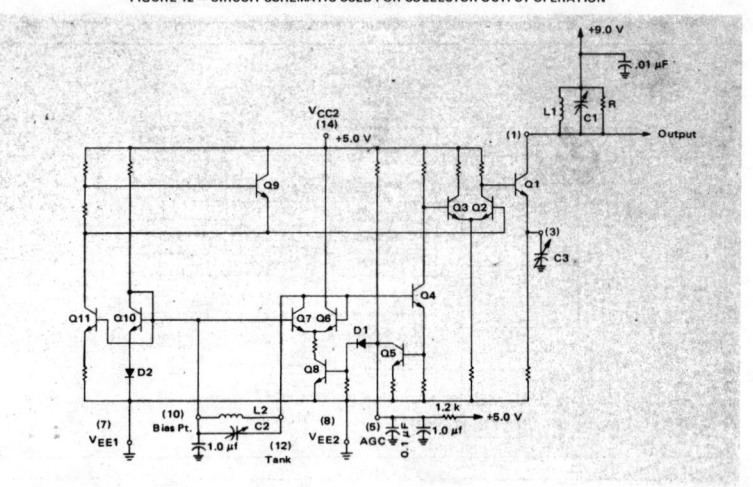

FIGURE 13 — POWER OUTPUT versus COLLECTOR LOAD

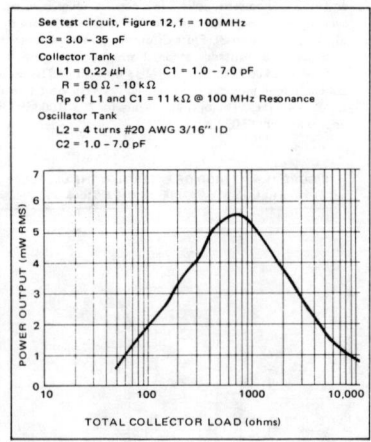

See test circuit, Figure 12, f = 100 MHz
C3 = 3.0 - 35 pF
Collector Tank
L1 = 0.22 µH   C1 = 1.0 - 7.0 pF
R = 50 Ω - 10 kΩ
R_p of L1 and C1 = 11 kΩ @ 100 MHz Resonance
Oscillator Tank
L2 = 4 turns #20 AWG 3/16" ID
C2 = 1.0 - 7.0 pF

FIGURE 14 — POWER OUTPUT versus COLLECTOR LOAD

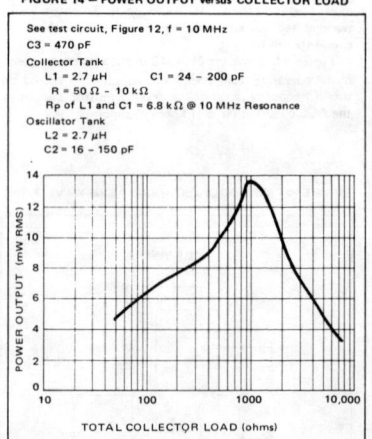

See test circuit, Figure 12, f = 10 MHz
C3 = 470 pF
Collector Tank
L1 = 2.7 µH   C1 = 24 - 200 pF
R = 50 Ω - 10 kΩ
R_p of L1 and C1 = 6.8 kΩ @ 10 MHz Resonance
Oscillator Tank
L2 = 2.7 µH
C2 = 16 - 150 pF

Circuit diagrams utilizing Motorola products are included as a means of illustrating typical semiconductor applications; consequently, complete information sufficient for construction purposes is not necessarily given. The information has been carefully checked and is believed to be entirely reliable. However, no responsibility is assumed for inaccuracies. Furthermore, such information does not convey to the purchaser of the semiconductor devices described any license under the patent rights of Motorola Inc. or others.

Courtesy Motorola Semiconductor Products, Inc.

MC4324 • MC4024

The MC4324/4024 consists of two independent voltage-controlled multivibrators with output buffers. Variation of the output frequency over a 3.5-to-1 range is guaranteed with an input dc control voltage of 1.0 to 5.0 volt.

Operating frequency is specified at 25 MHz at 25°C. Operation to 15 MHz is possible over the specified temperature range. For higher frequency requirements, see the MC1648 (200 MHz) or the MC1658 (125 MHz) data sheet.

This device was designed specifically for use in phase-locked loops for digital frequency control. It can also be used in other applications requiring a voltage-controlled frequency, or as a stable fixed frequency oscillator (3.0 MHz to 15 MHz) by replacing the external control capacitor with a crystal.

- Maximum Operating Frequency = 25 MHz Guaranteed @ 25°C
- Power Dissipation = 150 mW typ/pkg
- Output Loading Factor = 7

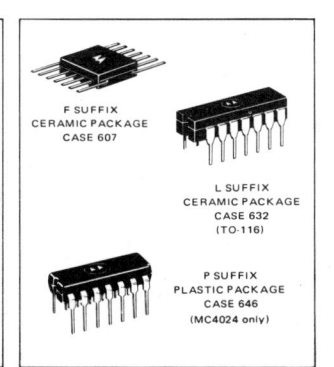

F SUFFIX
CERAMIC PACKAGE
CASE 607

L SUFFIX
CERAMIC PACKAGE
CASE 632
(TO-116)

P SUFFIX
PLASTIC PACKAGE
CASE 646
(MC4024 only)

TYPICAL APPLICATIONS

FIGURE 1 – ASTABLE MULTIVIBRATOR

f_{out} = 10 MHz

FIGURE 2 – CRYSTAL CONTROLLED MULTIVIBRATOR

Crystal frequency can be pulled slightly by adjusting P1.

FIGURE 3 – VOLTAGE-CONTROLLED MULTIVIBRATOR

V_{in} = 2.5 V to 5.5 V
f_{out} = 1.0 MHz min, 5.0 MHz max

FIGURE 4 – PHASE-LOCKED, FREQUENCY SYNTHESIZER LOOP

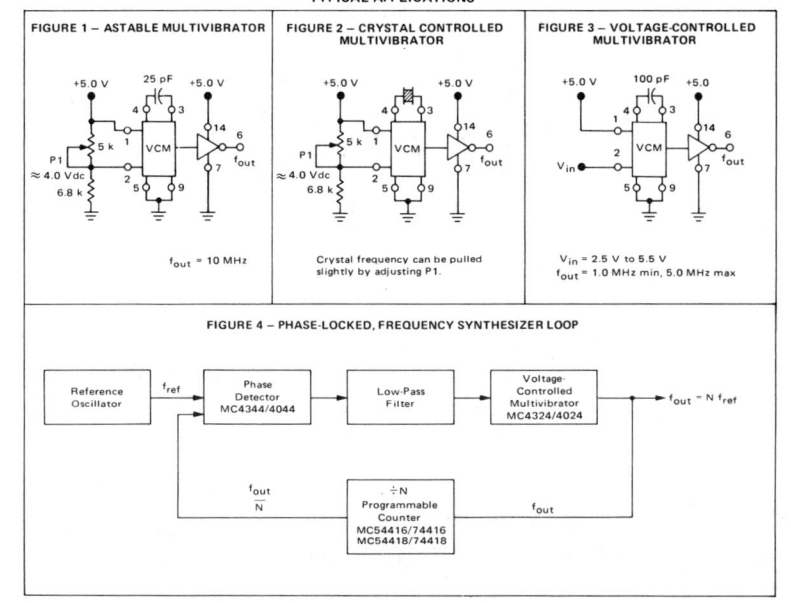

Reference Oscillator → f_{ref} → Phase Detector MC4344/4044 → Low-Pass Filter → Voltage-Controlled Multivibrator MC4324/4024 → f_{out} = N f_{ref}

÷ N Programmable Counter MC54416/74416 MC54418/74418 → f_{out}

Courtesy Motorola Semiconductor Products, Inc.

ELECTRICAL CHARACTERISTICS

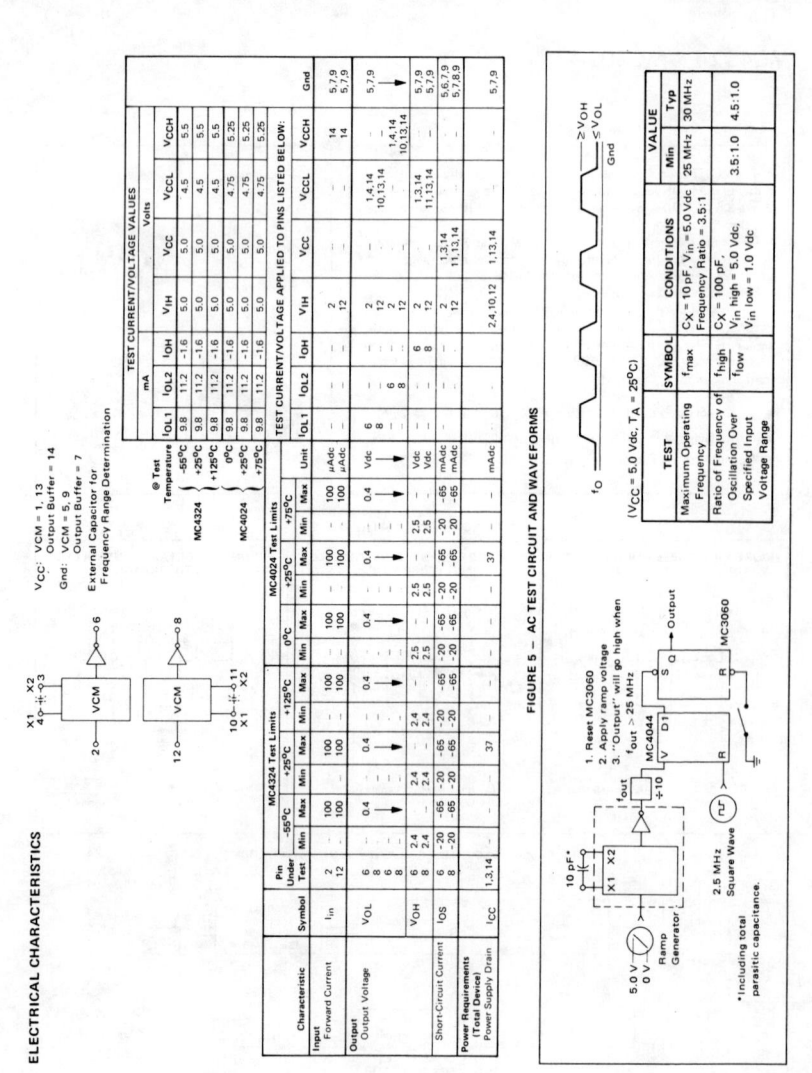

V_{CC}: VCM = 1, 13
Output Buffer = 14
Gnd: VCM = 5, 9
Output Buffer = 7
External Capacitor for
Frequency Range Determination

		Pin Under Test	MC4324 Test Limits −55°C Min	Max	MC4324 Test Limits +25°C Min	Max	MC4324 Test Limits +125°C Min	Max	MC4024 Test Limits 0°C Min	Max	MC4024 Test Limits +25°C Min	Max	MC4024 Test Limits +75°C Min	Max	Unit
Characteristic	Symbol														
Input Forward Current	I_{in}	2, 12	−	100	−	100	−	100	−	100	−	100	−	100	µAdc
Output Output Voltage	V_{OL}	6, 8	−	0.4	−	0.4	−	0.4	−	0.4	−	0.4	−	0.4	Vdc
	V_{OH}	6, 8	2.4	−	2.4	−	2.4	−	2.5	−	2.5	−	2.5	−	Vdc
Short-Circuit Current	I_{OS}	6, 8	−20	−65	−20	−65	−20	−65	−20	−65	−20	−65	−20	−65	mAdc
Power Requirements (Total Device) Power Supply Drain	I_{CC}	1,3,14	−	37							−	37			mAdc

TEST CURRENT/VOLTAGE VALUES

	mA				Volts			
@ Test Temperature	I_{OL1}	I_{OL2}	I_{OH}	V_{IH}	V_{CC}	V_{CCL}	V_{CCH}	
−55°C	9.8	11.2	−1.6	5.0	5.0	4.5	5.5	
+25°C	9.8	11.2	−1.6	5.0	5.0	4.5	5.5	
+125°C	9.8	11.2	−1.6	5.0	5.0	4.5	5.5	
0°C	9.8	11.2	−1.6	5.0	5.0	4.75	5.25	
+25°C	9.8	11.2	−1.6	5.0	5.0	4.75	5.25	
+75°C	9.8	11.2	−1.6	5.0	5.0	4.75	5.25	

TEST CURRENT/VOLTAGE APPLIED TO PINS LISTED BELOW:

	I_{OL1}	I_{OL2}	I_{OH}	V_{IH}	V_{CC}	V_{CCL}	V_{CCH}	Gnd
MC4324	−	−	6, 8	2, 12	−	1,4,14 10,13,14	1,4,14 10,13,14	5,7,9 5,7,9
MC4024	−	−	6, 8	2, 12	−	1,3,14 11,13,14	−	5,7,9 5,7,9
	2,4,10,12	1,13,14			1,3,14 11,13,14			5,6,7,9 5,7,8,9 5,7,9

FIGURE 5 − AC TEST CIRCUIT AND WAVEFORMS

(V_{CC} = 5.0 Vdc, T_A = 25°C)

TEST	SYMBOL	CONDITIONS	VALUE Min	VALUE Typ
Maximum Operating Frequency	f_{max}	C_X = 10 pF, V_{in} = 5.0 Vdc Frequency Ratio = 3.5:1	25 MHz	30 MHz
Ratio of Frequency of Oscillation Over Specified Input Voltage Range	f_{high}/f_{low}	C_X = 100 pF, V_{in} high = 5.0 Vdc, V_{in} low = 1.0 Vdc	3.5:1.0	4.5:1.0

Courtesy Motorola Semiconductor Products, Inc.

MAXIMUM RATINGS

Rating		Value	Unit
Supply Operating Voltage Range	MC4324	4.5 to 5.5	Vdc
	MC4024	4.75 to 5.25	
Supply Voltage		+7.0	Vdc
Input Voltage		+5.5	Vdc
Output Voltage		+5.5	Vdc
Operating Temperature Range	MC4324	−55 to +125	°C
	MC4024	0 to +75	
Storage Temperature Range — Ceramic Package		−65 to +150	°C
Plastic Package		−55 to +125	
Maximum Junction Temperature	MC4324	+175	°C
	MC4024	+150	
Thermal Resistance − Junction To Case (θ_{JC})			°C/mW
Flat Ceramic Package		0.06	
Dual In-Line Ceramic Package		0.05	
Plastic Package		0.07	
Thermal Resistance − Junction To Ambient (θ_{JA})			°C/mW
Flat Ceramic Package		0.21	
Dual In-Line Ceramic Package		0.15	
Plastic Package		0.15	

FIGURE 6 — CIRCUIT SCHEMATIC

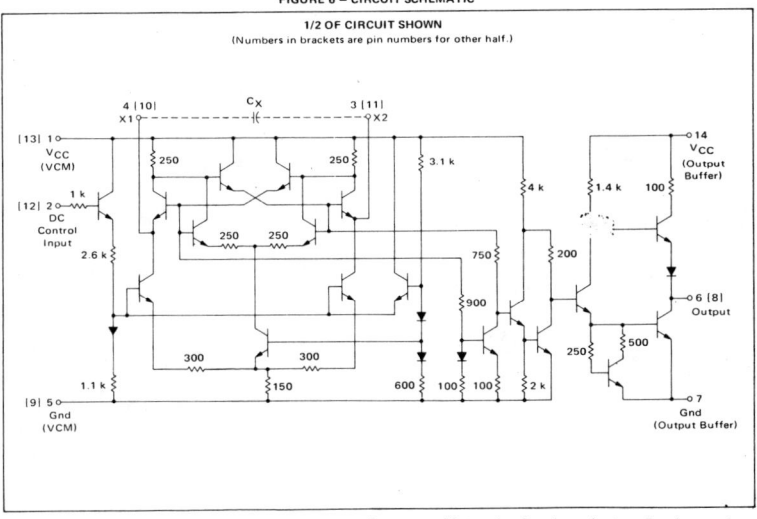

Courtesy Motorola Semiconductor Products, Inc.

FIGURE 7 — FREQUENCY-CAPACITANCE PRODUCT

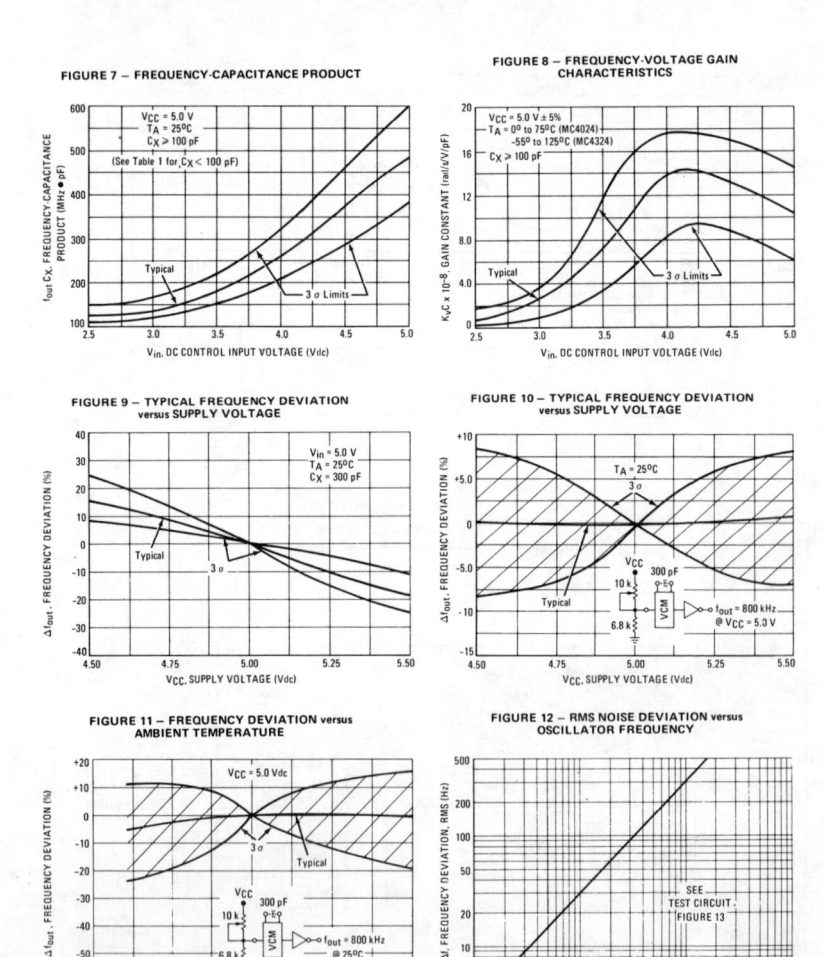

FIGURE 8 — FREQUENCY-VOLTAGE GAIN CHARACTERISTICS

FIGURE 9 — TYPICAL FREQUENCY DEVIATION versus SUPPLY VOLTAGE

FIGURE 10 — TYPICAL FREQUENCY DEVIATION versus SUPPLY VOLTAGE

FIGURE 11 — FREQUENCY DEVIATION versus AMBIENT TEMPERATURE

FIGURE 12 — RMS NOISE DEVIATION versus OSCILLATOR FREQUENCY

NOTE: Curves labeled as 3 σ limits denote that 99.7% of the devices tested fell within these limits.

Courtesy Motorola Semiconductor Products, Inc.

FIGURE 13 — NOISE DEVIATION TEST CIRCUIT

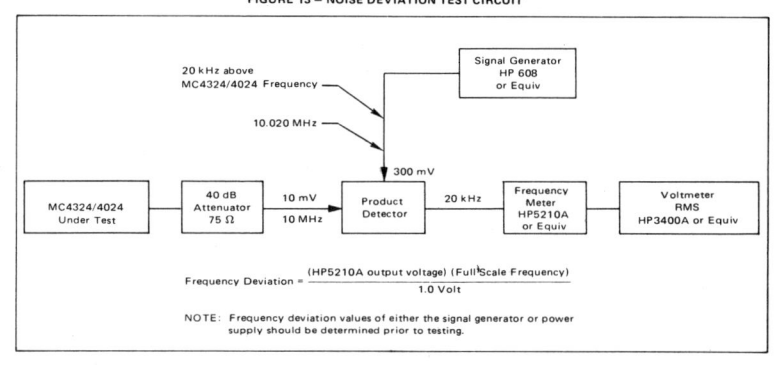

```
20 kHz above
MC4324/4024 Frequency ──┐      ┌─────────────────┐
                        │      │ Signal Generator│
         10.020 MHz ────┤      │     HP 608      │
                        │      │    or Equiv     │
                        ▼      └────────┬────────┘
                                        │
                                   300 mV
                                        ▼
┌────────────┐  ┌──────────┐  ┌──────┐  ┌──────────┐  ┌──────┐  ┌─────────────┐  ┌──────────────┐
│ MC4324/4024│──│  40 dB   │──│ 10 mV│──│ Product  │──│20 kHz│──│  Frequency  │──│  Voltmeter   │
│ Under Test │  │Attenuator│  │10 MHz│  │ Detector │  │      │  │    Meter    │  │     RMS      │
│            │  │   75 Ω   │  │      │  │          │  │      │  │   HP5210A   │  │HP3400A or    │
│            │  │          │  │      │  │          │  │      │  │   or Equiv  │  │    Equiv     │
└────────────┘  └──────────┘  └──────┘  └──────────┘  └──────┘  └─────────────┘  └──────────────┘
```

Frequency Deviation = $\dfrac{\text{(HP5210A output voltage)(Full Scale Frequency)}}{1.0 \text{ Volt}}$

NOTE: Frequency deviation values of either the signal generator or power supply should be determined prior to testing.

APPLICATIONS INFORMATION

Suggested Design Practices

Three power supply and three ground connections are provided in this circuit (each multivibrator has separate power supply and ground connections, and the output buffers have common power supply and ground pins). This provides isolation between VCM's and minimizes the effect of output buffer transients on the multivibrators in critical applications. The separation of power supply and ground lines also provides the capability of disabling one VCM by disconnecting its V_{CC} pin. However, all ground lines must always be connected to insure substrate grounding and proper isolation.

General design rules are:

1. Ground pins 5, 7, and 9 for all applications, including those where only one VCM is used.
2. Use capacitors with less than 50 nA leakage at plus and minus 3.0 volts. Capacitance values of 15 pF or greater are acceptable.
3. When operated in the free running mode, the minimum voltage applied to the DC Control input should be 60% of V_{CC} for good stability. The maximum voltage at this input should be V_{CC} + 0.5 volt.
4. When used in a phase-locked loop, the filter design should have a minimum DC Control input voltage of 1.0 volt and a maximum voltage of V_{CC} + 0.5 volt. The maximum restriction may be waived if the output impedance of the driving device is such that it will not source more than 10 mA at a voltage of V_{CC} + 0.5 volt.
5. The power supply for this device should be bypassed with a good quality RF-type capacitor of 500 to 1000 pF. Bypass capacitor lead lengths should be kept as short as possible. For best results, power supply voltage should be maintained as close to +5.0 V as possible. Under no conditions should the design require operation with a power supply voltage outside the range of 5.0 volts ± 10%.

External Control Capacitor (C_X) Determination (See Table 1)

The operating frequency range of this multivibrator is controlled by the value of an external capacitor that is connected between X1 and X2. A tuning ratio of 3.5-to-1 and a maximum frequency of 25 MHz are guaranteed under ideal conditions (V_{CC} = 5.0 volts, T_A = 25°C). Under actual operating conditions, variations in supply voltage, ambient temperature, and internal component tolerances limit the tuning ratio (see Figures 7 thru 12). An improvement in tuning ratio can be achieved by providing a variable tuning capacitor to facilitate initial alignment of the circuit.

Figures 7 through 11 show typical and suggested design limit information for important VCM characteristics. The suggested design limits are based on operation over the specified temperature range with a supply voltage of 5.0 volts ±5% unless otherwise noted. They include a safety factor of three times the estimated standard deviation.

Figures 7 and 8 provide data for any external control capacitor value greater than 100 pF. With smaller capacitor values, the curves are effectively moved downward. For example, a typical curve of frequency versus control voltage would be very nearly identical to the lower suggested

Courtesy Motorola Semiconductor Products, Inc.

TABLE 1 — EXTERNAL CONTROL CAPACITOR VALUE DETERMINATION

CONFIGURATION	T_A	V_{CC}	K1	K2	K3	K4	K5
With $C_X = \dfrac{K1}{f_{OH}} - 5$, $f_{OL} \approx \dfrac{K2}{C_X}$	25°C ±3°C	5.0 V	385	150	600	110	1.0
		5.0 V ±5%	325	175	680	125	1.14
		5.0 V ±10%	290	190	750	140	1.25
$C_X = C_{XV} + C_{XF}$ Choose C_{XF} and C_{XV} such that C_X can be adjusted to: $\dfrac{K1}{f_{OH}} - 5 \leqslant C_X \leqslant \dfrac{K3}{f_{OH}} - 5$ With $V_{in} = V_{CC} = 5.0$ V, adjust C_X to obtain: $f_{out} = K5\,(f_{OH})$ Then: $f_{OL} \approx \dfrac{K4}{K1} f_{OH}$	0°C to 75°C	5.0 V	335	165	660	120	1.10
		5.0 V ±5%	280	190	750	140	1.25
		5.0 V ±10%	250	200	840	150	1.40
	−55°C to 125°C	5.0 V	300	175	690	125	1.15
		5.0 V ±5%	260	200	780	145	1.30
		5.0 V ±10%	230	210	860	155	1.45

Definitions: f_{OH} = Output frequency with $V_{in} = V_{CC}$
f_{OL} = Output frequency with $V_{in} = 2.5$ V
(Frequencies in MHz, C_X in pF)

design limit of Figure 7 if a 15 pF capacitor is used. To use Figure 7, divide on the ordinate by the capacitor value in picofarads to obtain output frequency in megahertz. In Figure 8, the ordinate axis is multiplied by the capacitor value in picofarads to obtain the gain constant (K_V) in radians/second/volt.

Frequency Stability

When the MC4324/4024 is used as a fixed-frequency oscillator (V_{in} constant), the output frequency will vary slightly because of internal noise. This variation is indicated by Figure 12 for the circuit of Figure 13. These variations are relatively independent (< 10%) of changes in temperature and supply voltage.

10-to-1 Frequency Synthesizer

A frequency synthesizer covering a 10-to-1 range is shown in Figure 14. Three packages are required to complete the loop: The MC4344/4044 phase-frequency detector, the MC4324/4024 dual voltage-controlled multivibrator, and the MC54418/74418 programmable counter. Two VCM's (one package) are used to obtain the required frequency range. Each VCM is capable of operating over a 3-to-1 range, thus VCM1 is used for the lower portion of the times ten range and VCM2 covers the upper end. The proper divide ratio is set into the programmable counter and the VCM for that frequency is selected by control gates. The other VCM is left to be free running since its output is gated out of the feedback path.

Normally with a single VCM the loop gain would vary over a 10-to-1 range due to the range of the counter ratios. This affects the bandwidth, lockup time, and damping ratio severely. Utilizing two VCM's reduces this change in loop gain from 10-to-1 to 3-to-1 as a result of the different sensitivities of the two VCM's due to the different frequency ranges. This change of VCM sensitivity (3-to-1) is of such a direction to compensate for loop gain variations due to the programmable counter.

The overall concept of multi-VCM operation can be expanded for ranges greater than 10-to-1. Four VCM's (two packages) could be used to cover a 100-to-1 range.

Courtesy Motorola Semiconductor Products, Inc.

FIGURE 14 — 10-TO-1 FREQUENCY SYNTHESIZER

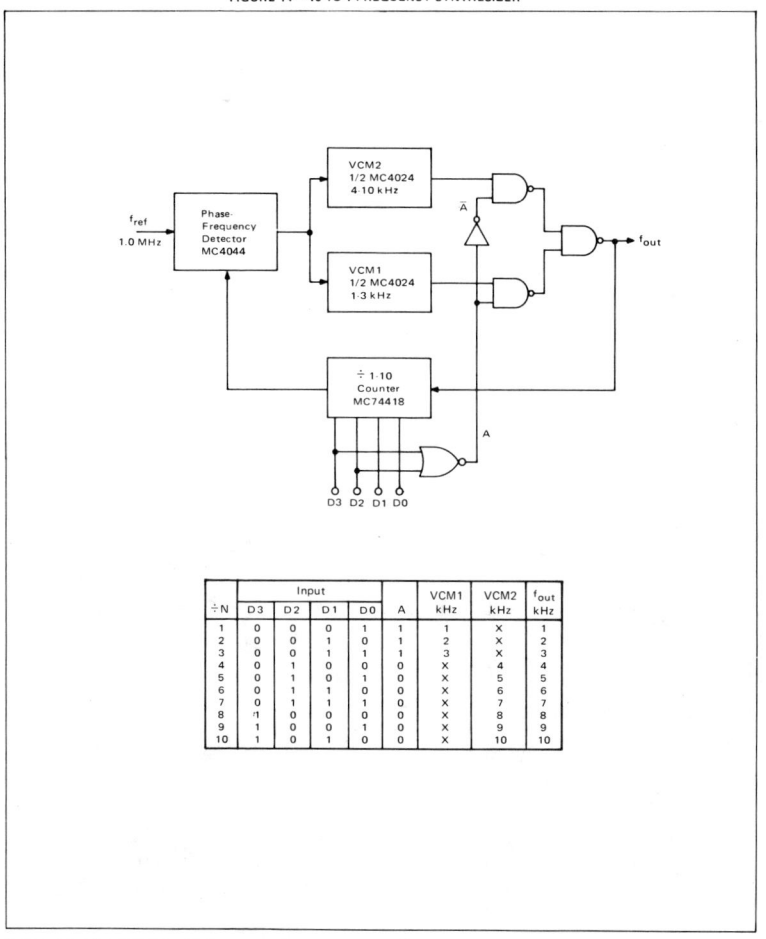

÷N	Input				A	VCM1 kHz	VCM2 kHz	f_{out} kHz
	D3	D2	D1	D0				
1	0	0	0	1	1	1	X	1
2	0	0	1	0	1	2	X	2
3	0	0	1	1	1	3	X	3
4	0	1	0	0	0	X	4	4
5	0	1	0	1	0	X	5	5
6	0	1	1	0	0	X	6	6
7	0	1	1	1	0	X	7	7
8	1	0	0	0	0	X	8	8
9	1	0	0	1	0	X	9	9
10	1	0	1	0	0	X	10	10

Circuit diagrams utilizing Motorola products are included as a means of illustrating typical semiconductor applications; consequently, complete information sufficient for construction purposes is not necessarily given. The information has been carefully checked and is believed to be entirely reliable. However, no responsibility is assumed for inaccuracies. Furthermore, such information does not convey to the purchaser of the semiconductor devices described any license under the patent rights of Motorola Inc. or others.

Courtesy Motorola Semiconductor Products, Inc.

CASE 607-05

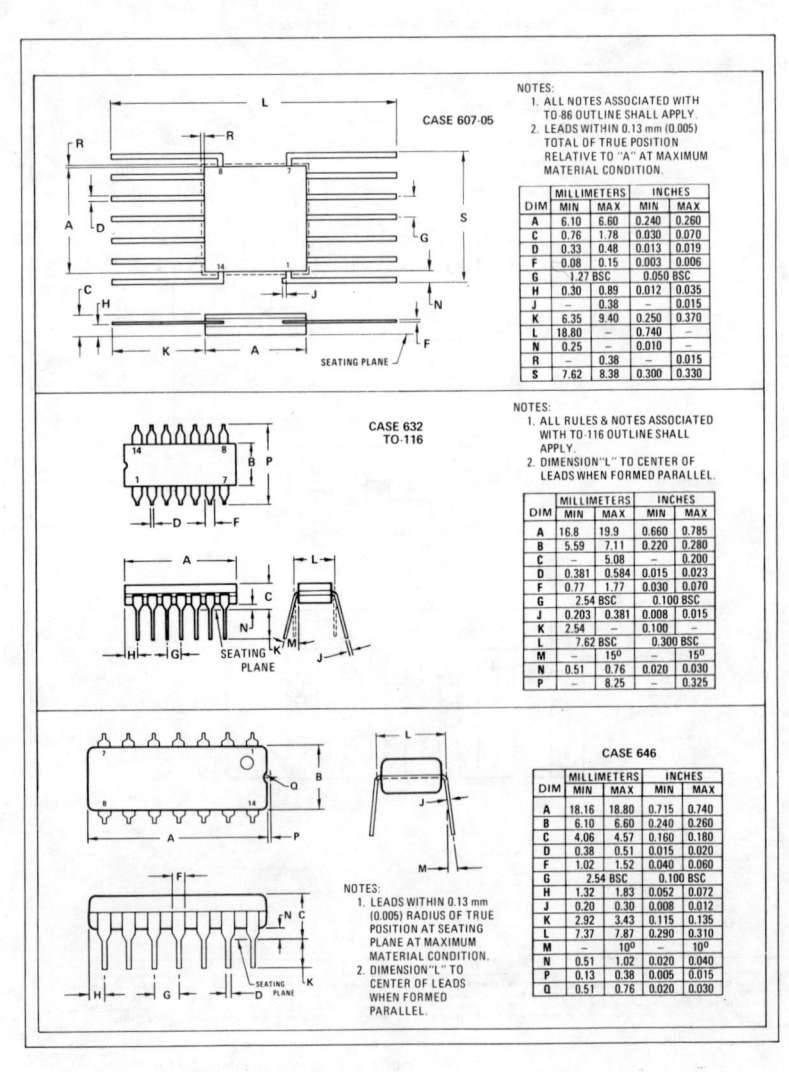

NOTES:
1. ALL NOTES ASSOCIATED WITH TO-86 OUTLINE SHALL APPLY.
2. LEADS WITHIN 0.13 mm (0.005) TOTAL OF TRUE POSITION RELATIVE TO "A" AT MAXIMUM MATERIAL CONDITION.

DIM	MILLIMETERS MIN	MAX	INCHES MIN	MAX
A	6.10	6.60	0.240	0.260
C	0.76	1.78	0.030	0.070
D	0.33	0.48	0.013	0.019
F	0.08	0.15	0.003	0.006
G	1.27 BSC		0.050 BSC	
H	0.30	0.89	0.012	0.035
J	–	0.38	–	0.015
K	6.35	9.40	0.250	0.370
L	18.80	–	0.740	–
N	0.25	–	0.010	–
R	–	0.38	–	0.015
S	7.62	8.38	0.300	0.330

CASE 632 TO-116

NOTES:
1. ALL RULES & NOTES ASSOCIATED WITH TO-116 OUTLINE SHALL APPLY.
2. DIMENSION "L" TO CENTER OF LEADS WHEN FORMED PARALLEL.

DIM	MILLIMETERS MIN	MAX	INCHES MIN	MAX
A	16.8	19.9	0.660	0.785
B	5.59	7.11	0.220	0.280
C	–	5.08	–	0.200
D	0.381	0.584	0.015	0.023
F	0.77	1.77	0.030	0.070
G	2.54 BSC		0.100 BSC	
J	0.203	0.381	0.008	0.015
K	2.54	–	0.100	–
L	7.62 BSC		0.300 BSC	
M	–	15°	–	15°
N	0.51	0.76	0.020	0.030
P	–	8.25	–	0.325

CASE 646

NOTES:
1. LEADS WITHIN 0.13 mm (0.005) RADIUS OF TRUE POSITION AT SEATING PLANE AT MAXIMUM MATERIAL CONDITION.
2. DIMENSION "L" TO CENTER OF LEADS WHEN FORMED PARALLEL.

DIM	MILLIMETERS MIN	MAX	INCHES MIN	MAX
A	18.16	18.80	0.715	0.740
B	6.10	6.60	0.240	0.260
C	4.06	4.57	0.160	0.180
D	0.38	0.51	0.015	0.020
F	1.02	1.52	0.040	0.060
G	2.54 BSC		0.100 BSC	
H	1.32	1.83	0.052	0.072
J	0.20	0.30	0.008	0.012
K	2.92	3.43	0.115	0.135
L	7.37	7.87	0.290	0.310
M	–	10°	–	10°
N	0.51	1.02	0.020	0.040
P	0.13	0.38	0.005	0.015
Q	0.51	0.76	0.020	0.030

Courtesy Motorola Semiconductor Products, Inc.

MC4344 • MC4044

The MC4344/4044 consists of two digital phase detectors, a charge pump, and an amplifier. In combination with a voltage controlled multivibrator (such as the MC4324/4024 or MC1648), it is useful in a broad range of phase-locked loop applications. The circuit accepts MTTL waveforms at the R and V inputs and generates an error voltage that is proportional to the frequency and/or phase difference of the input signals. Phase detector #1 is intended for use in systems requiring zero frequency and phase difference at lock. Phase detector #2 is used if quadrature lock is desired. Phase detector #2 can also be used to indicate that the main loop, utilizing phase detector #1, is out of lock.

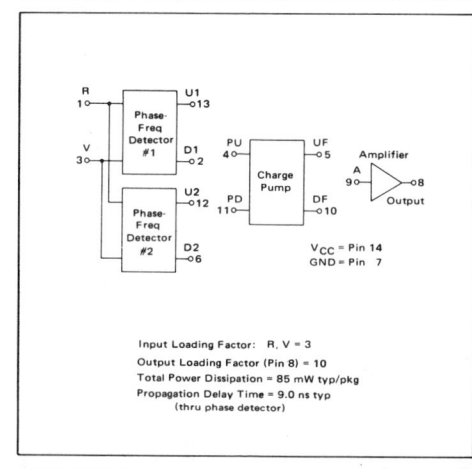

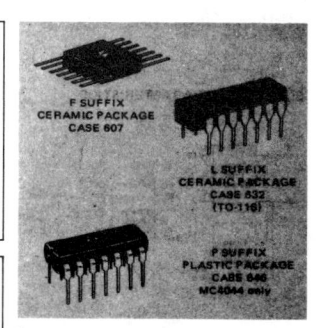

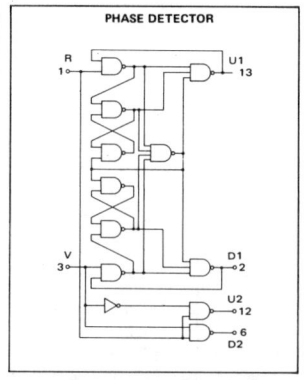

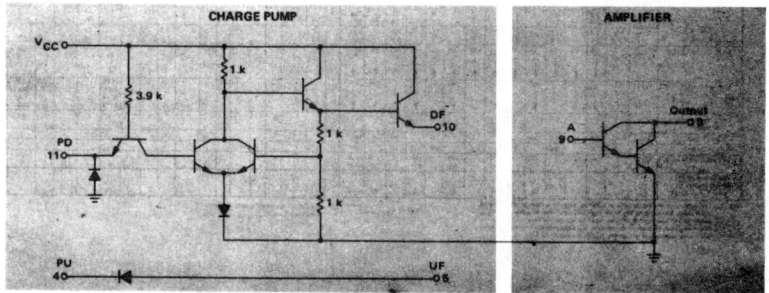

Input Loading Factor: R, V = 3
Output Loading Factor (Pin 8) = 10
Total Power Dissipation = 85 mW typ/pkg
Propagation Delay Time = 9.0 ns typ
(thru phase detector)

V_{CC} = Pin 14
GND = Pin 7

Courtesy Motorola Semiconductor Products, Inc.

217

ELECTRICAL CHARACTERISTICS

TRUTH TABLE

This is not strictly a functional truth table; i.e., it does not show all possible modes of operation. It is useful for dc testing.

1. X indicates output state unknown.
2. U1 and D1 outputs are sequential; i.e., they must be sequenced in order shown.
3. U2 and D2 outputs are combinational; i.e., they need only inputs shown to obtain outputs.

Courtesy Motorola Semiconductor Products, Inc.

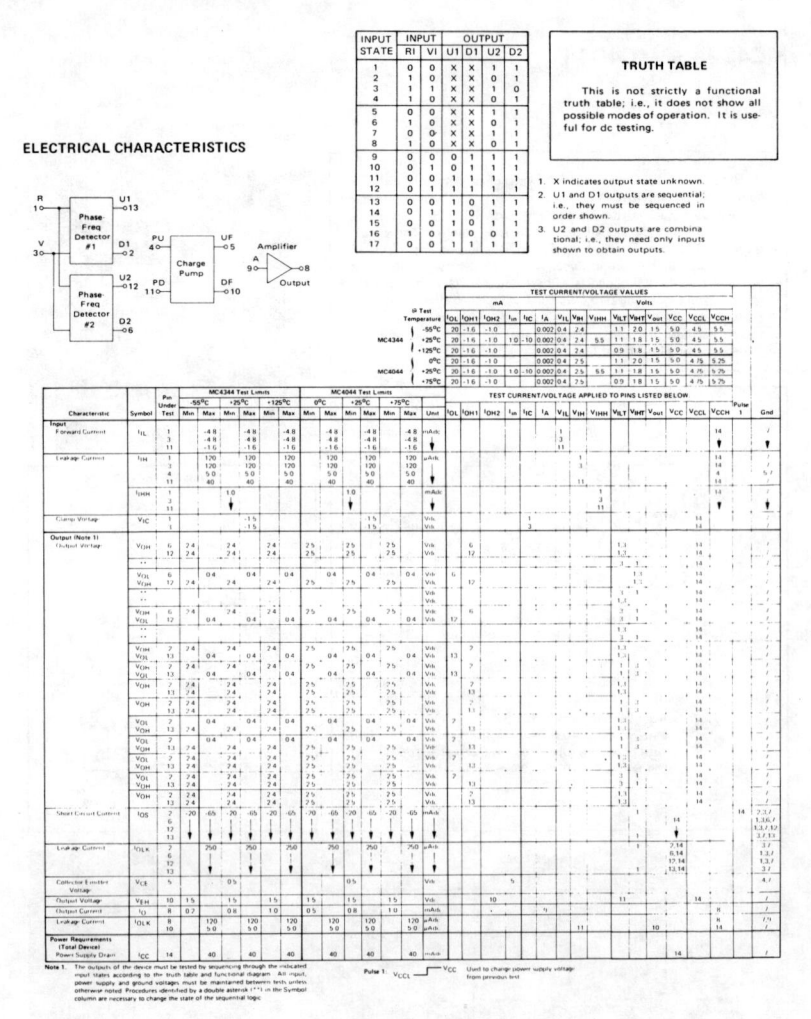

218

MAXIMUM RATINGS

Rating		Value	Unit
Supply Operating Voltage Range	MC4344	4.5 to 5.5	Vdc
	MC4044	4.75 to 5.25	
Supply Voltage		+7.0	Vdc
Input Voltage		+5.5	Vdc
Output Voltage		+5.5	Vdc
Operating Temperature Range	MC4344	-55 to +125	°C
	MC4044	0 to +75	
Storage Temperature Range — Ceramic Package		-65 to +150	°C
Plastic Package		-55 to +125	
Maximum Junction Temperature	MC4344	+175	°C
	MC4044	+150	
Thermal Resistance – Junction To Case (θ_{JC})			°C/mW
Flat Ceramic Package		0.06	
Dual In-Line Ceramic Package		0.05	
Plastic Package		0.07	
Thermal Resistance - Junction To Ambient (θ_{JA})			°C/mW
Flat Ceramic Package		0.21	
Dual In-Line Ceramic Package		0.15	
Plastic Package		0.15	

CONTENTS

	Page		Page
Operating Characteristics	3	Spurious Outputs	10
Phase-Locked Loop Components	6	Additional Loop Filtering	11
General	6	Applications Information	14
Loop Filter	7	Frequency Synthesizers	14
Design Problems and Their Solutions	9	Clock Recovery from Phase-Encoded Data	16
Dynamic Range	9	Package Dimensions	20

OPERATING CHARACTERISTICS

Operation of the MC4344/4044 is best explained by initially considering each section separately. If phase detector #1 is used, loop lockup occurs when both outputs U1 and D1 remain high. This occurs only when all the negative transitions on R, the reference input, and V, the variable or feedback input, coincide. The circuit responds only to transitions, hence phase error is independent of input waveform duty cycle or amplitude variation. Phase detector #1 consists of sequential logic circuitry, therefore operation prior to lockup is determined by initial conditions.

When operation is initiated, by either applying power to the circuit or active input signals to R and V, the circuitry can be in one of several states. Given any particular starting conditions, the flow table of Figure 1 can be used to determine subsequent operation. The flow table indicates the status of U1 and D1 as the R and V inputs are varied. The numbers in the table which are in parentheses are arbitrarily assigned labels that correspond to stable states that can result for each input combination. The numbers without parentheses refer to unstable conditions. Input changes are traced by horizontal movement in the table; after each input change, circuit operation will settle in the numbered state indicated by moving horizontally to the appropriate R-V column. If the number at that

FIGURE 1 – PHASE DETECTOR #1 FLOW TABLE

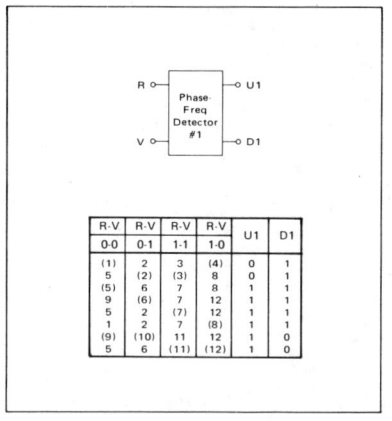

R-V	R-V	R-V	R-V	U1	D1
0-0	0-1	1-1	1-0		
(1)	2	3	(4)	0	1
5	(2)	(3)	8	0	1
(5)	6	7	8	1	1
9	(6)	7	12	1	1
5	2	(7)	12	1	1
1	2	7	(8)	1	1
(9)	(10)	11	12	1	0
5	6	(11)	(12)	1	0

Courtesy Motorola Semiconductor Products, Inc.

219

FIGURE 2 – PHASE DETECTOR #1 TIMING DIAGRAM

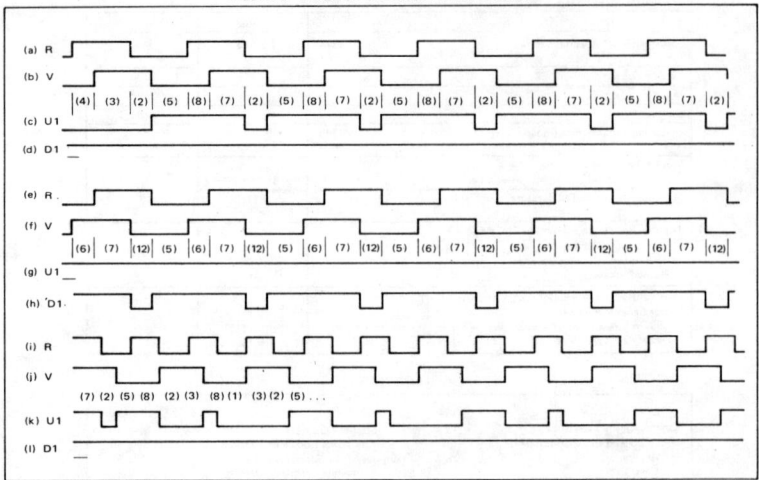

location is not in parentheses, move vertically to the number of the same value that is in parentheses. For a given input pair, any one of three stable states can exist. As an example, if R = 1 and V = 0, the circuit will be in one of the stable states (4), (8), or (12).

Use of the table in determining circuit operation is illustrated in Figure 2. In the timing diagram, the input to R is the reference frequency; the input to V is the same frequency but lags in phase. Stable state (4) is arbitrarily assumed as the initial condition. From the timing diagram and flow table, when the circuit is in stable state (4), outputs U1 and D1 are "0" and "1" respectively. The next input state is R·V = 1·1; moving horizontally from stable state (4) under R·V = 1·0 to the R·V = 1·1 column, state 3 is indicated. However, this is an unstable condition and the circuit will assume the state indicated by moving vertically in the R·V = 1·1 column to stable state (3). In this instance, outputs U1 and D1 remain unchanged. The input states next become R·V = 0·1; moving horizontally to the R·V = 0·1 column, stable state (2) is indicated. At this point there is still no change in U1 or D1. The next input change shifts operation to the R·V = 0·0 column where unstable state 5 is indicated. Moving vertically to stable state (5), the outputs now change state to U1·D1 = 1·1. The next input change, R·V = 1·0, drives the circuitry to stable state (8), with no change in U1 or D1. The next input, R·V = 1·1, leads to stable state (7) with no change in the outputs. The next two input state changes cause U1 to go low between the negative transitions of R and V. As the inputs continue to change, the circuitry moves repeatedly through stable states (2), (5), (8), (7), (2), etc. as shown, and a periodic waveform is obtained on the U1 terminal while D1 remains high.

A similar result is obtained if V is leading with respect to R, except that the periodic waveform now appears on D1 as shown in rows e-h of the timing diagram of Figure 2. In each case, the average value of the resulting waveform is proportional to the phase difference between the two inputs. In a closed loop application, the error signal for controlling the VCO is derived by translating and filtering these waveforms.

The results obtained when R and V are separated by a fixed frequency difference are indicated in rows i-l of the timing system. For this case, the U1 output goes low when R goes low and stays in that state until a negative transition on V occurs. The resulting waveform is similar to the fixed phase difference case, but now the duty cycle of the U1 waveform varies at a rate proportional to the difference frequency of the two inputs, R and V. It is this characteristic that permits the MC4344/4044 to be used as a frequency discriminator; if the signal on R has been frequency modulated and if the loop bandwidth is selected to pass the deviation frequency but reject R and V, the resulting error voltage applied to the VCO will be the recovered modulation signal.

Phase detector #2 consists only of combinatorial logic, therefore its characteristics can be determined from the

Courtesy Motorola Semiconductor Products, Inc.

FIGURE 3 — PHASE DETECTOR #2 OPERATION

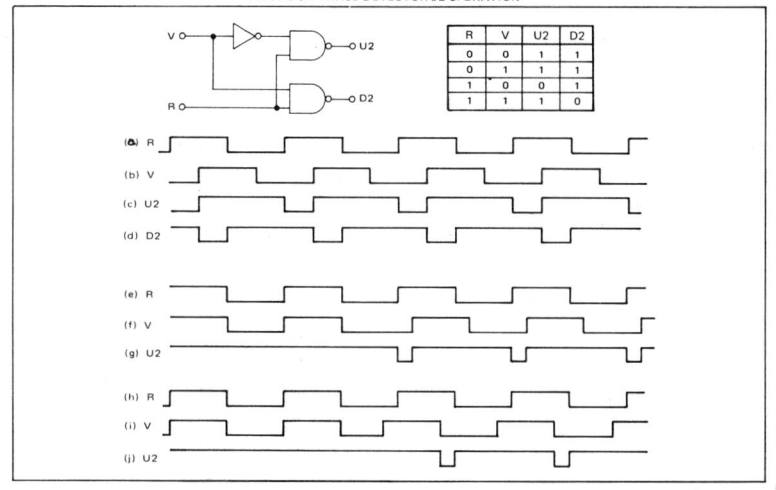

R	V	U2	D2
0	0	1	1
0	1	1	1
1	0	0	1
1	1	1	0

simple truth table of Figure 3. Since circuit operation requires that both inputs to the charge pump either be high or have the same duty cycle when lock occurs, using this phase detector leads to a quadrature relationship between R and V. This is illustrated in rows a-d of the timing diagram of Figure 3. Note that any deviation from a fifty percent duty cycle on the inputs would appear as phase error.

Waveforms showing the operation of phase detector #2 when phase detector #1 is being used in a closed loop are indicated in rows e-j. When the main loop is locked, U2 remains high. If the loop drifts out of lock in either direction a negative pulse whose width is proportional to the amount of drift appears on U2. This can be used to generate a simple loss-of-lock indicator.

Operation of the charge pump is best explained by considering it in conjunction with the Darlington amplifier included in the package (see Figure 4). There will be a pulsed waveform on either PD or PU, depending on the phase-frequency relationship of R and V. The charge pump serves to invert one of the input waveforms (D1) and translates the voltage levels before they are applied to the loop filter. When PD is low and PU is high, Q1 will be conducting in the normal direction and Q2 will be off. Current will be flowing through Q3 and CR2; the base of Q3 will be two V_{BE} drops above ground or approximately 1.5 volts. Since both of the resistors connected to the base of Q3 are equal, the emitter of Q4 (base of Q5) will be

FIGURE 4 — CHARGE PUMP OPERATION

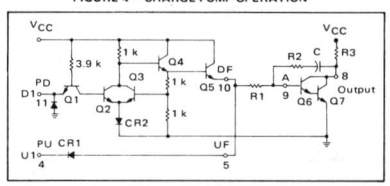

approximately 3.0 volts. For this condition, the emitter of Q5 (DF) will be one V_{BE} below this voltage, or about 2.25 volts. The PU input to the charge pump is high (> 2.4 volts) and CR1 will be reverse biased. Therefore Q5 will be supplying current to Q6. This will tend to lower the voltage at the collector of Q7, resulting in an error signal that lowers the VCO frequency as required by a "pump down" signal.

When PU is low and PD is high, CR1 is forward biased and UF will be approximately one V_{BE} above ground (neglecting the $V_{CE(sat)}$ of the driving gate). With PD high, Q1 conducts in the reverse direction, supplying base current for Q2. While Q2 is conducting, Q4 is prevented from supplying base drive to Q5; with Q5 cut off and UF low there is no base current for Q6 and the voltage at the

Courtesy Motorola Semiconductor Products, Inc.

collector of Q7 moves up, resulting in an increase in the VCO operating frequency as required by a "pump up" signal.

If both inputs to the charge pump are high (zero phase difference), both CR1 and the base-emitter junction of Q5 are reverse biased and there is no tendency for the error voltage to change. The output of the charge pump varies between one V_{BE} and three V_{BE} as the phase difference of R and V varies from minus 2π to plus 2π. If this signal is filtered to remove the high-frequency components, the phase detector transfer function, K_ϕ, of approximately 0.12 volt/radian is obtained (see Figure 5).

FIGURE 5 – PHASE DETECTOR TEST

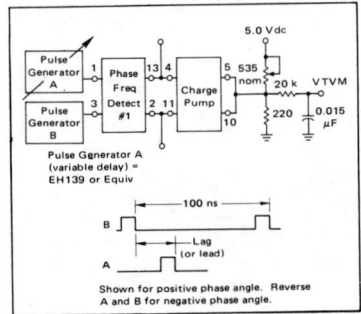

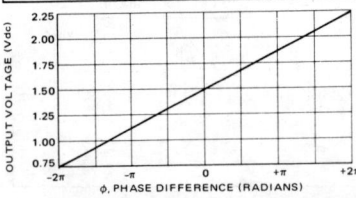

The specified gain constant of 0.12 volt/radian may not be obtained if the amplifier/filter combination is improperly designed. As indicated previously, the charge pump delivers pump commands of about 2.25 volts on the positive swings and 0.75 volt on the negative swings for a mean no-pump value of 1.5 volts. If the filter amplifier is biased to threshold "on" at 1.5 volts, then the pump up and down voltages have equal effects. The pump signals are established by V_{BE}'s of transistors with milliamperes of current flowing. On the other hand, the transistors included for use as a filter amplifier will have very small currents flowing and will have correspondingly lower V_{BE}'s — on the order of 0.6 volt each for a threshold of 1.2 volts. Any displacement of the threshold from 1.5 volts causes an increase in gain in one direction and a reduction in the other. The transistor configuration provided is hence not optimum but does allow for the use of an additional transistor to improve filter response. This addition also results in a non-symmetrical response since the threshold is now approximately 1.8 volts. The effective positive swing is limited to 0.45 volt while the negative swing below threshold can be greater than 1.0 volt. This means that the loop gain when changing from a high frequency to a lower frequency is less than when changing in the opposite direction. For type two loops this tends to increase overshoot when going from low to high and increases damping in the other direction. These problems and the selection of external filter components are intimately related to system requirements and are discussed in detail in the filter design section.

PHASE-LOCKED LOOP COMPONENTS
General

A basic phase-locked loop, when operating properly, will acquire ("lock on") an input signal, track it in frequency, and exhibit a fixed phase relationship relative to the input. In this basic loop, the output frequency will be identical to the input frequency (Figure 6). A fundamental loop consists of a phase detector, amplifier/filter, and voltage-controlled oscillator (Figure 7). It appears and acts like a unity gain feedback loop. The controlled variable is phase; any error between f_{in} and f_{out} is amplified and applied to the VCO in a corrective direction.

Simple phase detectors in digital phase-locked loops usually put out a series of pulses. The average value of these pulses is the "gain constant", K_ϕ, of the phase detector — the volts out for a given phase difference, expressed as volts/radian.

The VCO is designed so that its output frequency range is equal to or greater than the required output frequency range of the system. The ratio of change in output frequency to input control voltage is called "gain constant", K_O. If the slope of f_{out} to V_{in} is not linear (i.e., changes greater than 25%) over the expected frequency range, the

FIGURE 6 – BASIC PHASE-LOCKED LOOP FREQUENCY RELATIONSHIP

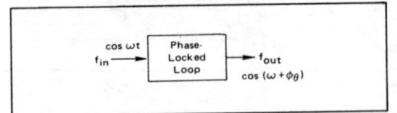

FIGURE 7 – FUNDAMENTAL PHASE-LOCKED LOOP

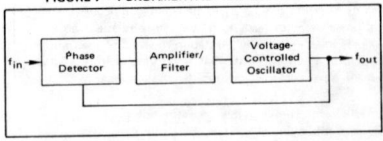

Courtesy Motorola Semiconductor Products, Inc.

curve should be piece-wise approximated and the appropriate constant applied for "best" and "worst" case analysis of loop performance.

System dynamics when in lock are determined by the amplifier/filter block. Its gain determines how much phase error exists between f_{in} and f_{out}, and filter characteristics shape the capture range and transient performance. This will be discussed in detail later.

Loop Filter

Fundamental loop characteristics such as capture range, loop bandwidth, capture time, and transient response are controlled primarily by the loop filter. The loop behavior is described by gains in each component block of Figure 8.

FIGURE 8 — GAIN CONSTANTS

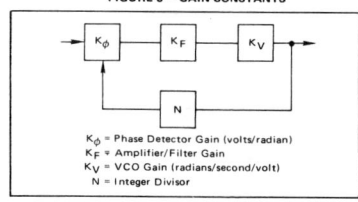

K_ϕ = Phase Detector Gain (volts/radian)
K_F = Amplifier/Filter Gain
K_V = VCO Gain (radians/second/volt)
N = Integer Divisor

The output to input ratio reflects a second order low pass filter in frequency response with a static gain of N:

$$\frac{\theta_O(s)}{\theta_i(s)} = \frac{K_\phi K_F K_V}{s + \frac{K_\phi K_F K_V}{N}} \quad (1)$$

where: $K_F = \frac{1 + T_1 s}{T_2 s}$ (2)

$T_1 = R_2 C$ and $T_2 = R_1 C$ of Figure 4. Therefore,

$$\frac{\theta_O(s)}{\theta_i(s)} = \frac{N(1 + T_1 s)}{\frac{s^2 N T_2}{K_\phi K_V} + T_1 s + 1} \quad (3)$$

Both ω_n (loop bandwidth or natural frequency) and ζ (damping factor) are particularly important in the transient response to a step input of phase or frequency (Figure 9), and are defined as:

$$\omega_n = \sqrt{\frac{K_\phi K_V}{N T_2}} \quad (4)$$

$$\zeta = \sqrt{\frac{K_\phi K_V}{N T_2}} \left(\frac{T_1}{2}\right) \quad (5)$$

Using these terms in Equation 3,

$$\frac{\theta_O(s)}{\theta_i(s)} = \frac{N(1 + T_1 s)}{\frac{s^2}{\omega_n^2} + \frac{2\zeta s}{\omega_n} + 1} \quad (6)$$

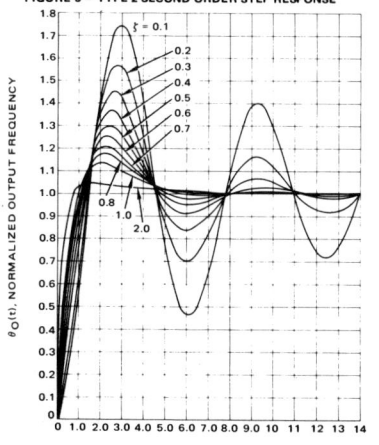

FIGURE 9 — TYPE 2 SECOND ORDER STEP RESPONSE

In a well defined system controlling factors such as ω_n and ζ may be chosen either from a transient basis (time domain response) or steady state frequency plot (roll-off point and peaking versus frequency). Once these two design goals are defined, synthesis of the filter is relatively straight-forward.

Constants K_ϕ, K_V, and N are usually fixed due to other design constraints, leaving T_1 and T_2 as variables to set ω_n and ζ. Since only T_2 appears in Equation 4, it is the easiest to solve for initially.

$$T_2 = \frac{K_\phi K_V}{N \omega_n^2} \quad (7)$$

From Equation 5, we find

$$T_1 = \frac{2\zeta}{\omega_n} \quad (8)$$

Using relationships 7 and 8, actual resistor values may be computed:

$$R_1 = \frac{K_\phi K_V}{N \omega_n^2 C} \quad (9)$$

$$R_2 = \frac{2}{\omega_n C} \quad (10)$$

Courtesy Motorola Semiconductor Products, Inc.

Although fundamentally the range of R_1 and R_2 may be from several hundred to several thousand ohms, sideband considerations usually force the value of R_1 to be set first, and then R_2 and C computed.

$$C = \frac{K_\phi K_V}{N\omega_n^2 R_1} \qquad (11)$$

Calculation of passive components R_2 and C (in synthesizers) is complicated by incomplete information on N, which is variable, and the limits of ω_n and ζ during that variance. Equally important are changes in K_V over the output frequency range. Minimum and maximum values of ω_n and ζ can be computed from Equations 4 and 5 when the appropriate worst case numbers are known for all the factors.

Amplifier/filter gain usually determines how much phase error exists between f_{in} and f_{out}, and the filter characteristic shapes capture range and transient performance. A relatively simple, low gain amplifier may usually be used in the loop since many designs are not constrained so much by phase error as by the need to make f_{in} equal f_{out}. Unnecessarily high gains can cause problems in linear loops when the system is out of lock if the amplifier output swing is not adequately restricted since integrating operational amplifier circuits will latch up in time and effectively open the loop.

The internal amplifier included in the MC4344/4044 may be used effectively if its limits are observed. The circuit configuration shown in Figure 10 illustrates the

FIGURE 10 — USING MC4344/4044 LOOP AMPLIFIER

placement of R_1, R_2, C, and load resistor R_L (1 kΩ). Due to the non-infinite gain of this stage ($A_V \approx 30$) and other non-ideal characteristics, some restraint must be placed on passive component selection. Foremost is a lower limit on the value of R_2 and an upper limit on R_1. Placed in order of priority, the recommendations are as follows: (a) $R_2 > 50\,\Omega$, (b) $R_2/R_1 \leq 10$, (c) $1\,k\Omega < R_1 < 5\,k\Omega$.

Limit (c) is the most flexible and may be violated with either higher sidebands and phase error ($R_1 > 5$ kΩ) or lower phase detector gain ($R_1 < 1$ kΩ). If limit (b) is exceeded, loop bandwidth will be less than computed and may not have any similarity to the prediction. For an accurate reproduction of calculated loop characteristics one should go to an operational amplifier which has sufficient gain to make limit (b) readily satisfied. Limit (a)

is very important because T_1 in Equation 5 is in reality composed of three elements:

$$T_1 = C \left(R_2 - \frac{1}{g_m} \right) \qquad (12)$$

where g_m = transconductance of the common emitter amplifier.

Normally g_m is large and T_1 nearly equals R_2C, but resistance values below 50 Ω can force the phase-compensating "zero" to infinity or worse (into the right half plane) and give an unstable system. The problem can be circumvented to a large degree by buffering the feedback with an emitter follower (Figure 11). Inequality (a) may

FIGURE 11 — AMPLIFIER CAPABLE OF HANDLING LOWER R2

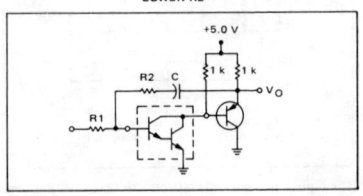

then be reduced by at least an order of magnitude ($R_2 > 5\,\Omega$) keeping in mind that electrolytic capacitors used as C may approach this value by themselves at the frequency of interest (ω_n).

Larger values of R_1 may be accommodated by either using an operational amplifier with a low bias current ($I_b < 1.0\,\mu A$) as shown in Figure 12 or by buffering the internal

FIGURE 12 — USING AN OPERATIONAL AMPLIFIER TO EXTEND THE VALUE OF R1

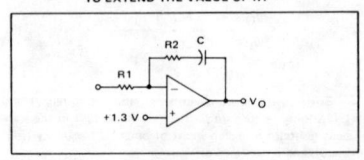

Darlington pair with an FET (Figure 13). It is vitally important, however, that the added device be operated at zero V_{GS}. Source resistor R_4 should be adjusted for this condition (which amounts to I_{DSS} current for the FET). This insures that the overall amplifier input threshold remains at the proper potential of approximately two base-emitter drops. Use of an additional emitter follower instead of the FET and R_4 (Figure 14) gives a threshold near the upper limit of the phase detector charge pump, resulting in an extremely unsymmetrical phase detector gain in the pump up versus pump down mode. It is not unusual to

Courtesy Motorola Semiconductor Products, Inc.

FIGURE 13 — FET BUFFERING TO RAISE AMPLIFIER INPUT IMPEDANCE

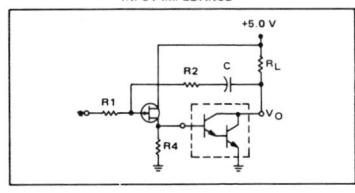

FIGURE 14 — EMITTER FOLLOWER BUFFERING OF AMPLIFIER INPUT

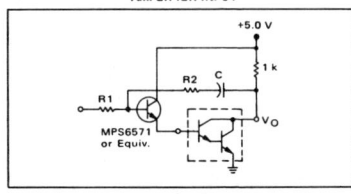

note a 5:1 difference in K_ϕ for circuits having the bipolar buffer stage. If the initial design can withstand this variation in loop gain and remain stable, the approach should be considered since there are no critical adjustments as in the FET circuit.

DESIGN PROBLEMS AND THEIR SOLUTIONS

Dynamic Range

A source of trouble for all phase-locked loops, as well as most electronics is simply overload or lack of sufficient dynamic range. One limit is the amplifier output drive to the VCO. Not only must a designer note the outside limits of the dc control voltage necessary to give the output frequency range, he must also account for the worst case of overshoot expected for the system. Relatively large damping factors ($\zeta = 0.5$) can contribute significant amounts of overshoot (30%). To be prepared for the worst case output swing the amplifier should have as much margin to positive and negative limits as the expected swing itself. That is, if a two-volt swing is sufficient to give the desired output frequency excursion, there should be at least a two-volt cushion above and below maximum expected steady-state values on the control line.

This increase in range, in order to be effective, must of course be followed by an equivalent range in the VCO or there is little to be gained. Any loss in loop gain will in general cause a decrease in ζ and a consequent increase in overshoot and ringing. If the loss in gain is caused by saturation or near saturation conditions, the problem tends to accelerate towards a situation where the system settles in not only a slow but oscillatory manner as well.

Loss of amplifier gain may not be due entirely to normal system damping considerations. In loops employing digital phase detectors, an additional problem is likely to appear. This is due to amplifier saturation during a step input when there is a maximum phase detector output simultaneous with a large transient overshoot. The phase detector square wave rides on top of the normal transient and may even exceed the amplifier output limits imposed above. Since the input frequency will exceed the R_2C time constant, gain K_F for these annoying pulses will be R_2/R_1. Ordinarily this ratio will be less than 1, but some circumstances dictate a low loop gain commensurate with a fairly high ω_n. For these cases, R_2/R_1 may be higher than 10 and cause pulse-wise saturation of the amplifier. Since the dc control voltage is an average of phase detector pulses, clipping can be translated into a reduction in gain with all the "benefits" already outlined, i.e., poor settling time. An easy remedy to apply in many cases is a simple RC low pass section preceding or together with the integrator-lag section. To make transient suppression independent of amplifier response, the network may be imbedded within the input resistor R_1 (Figure 15) or be implemented

FIGURE 15 — IMPROVED TRANSIENT SUPPRESSION WITH $R_1 - C_c$

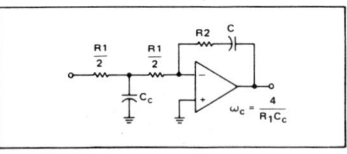

$$\omega_c = \frac{4}{R_1 C_c}$$

FIGURE 16 — IMPROVED TRANSIENT SUPPRESSION WITH $R_2 - C_c$

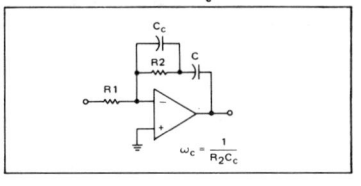

$$\omega_c = \frac{1}{R_2 C_c}$$

by placing a feedback capacitor across R_2 (Figure 16). Besides rounding off and inhibiting pulses, these networks add an additional pole to the loop and may cause further overshoot if the cutoff frequency (ω_c) is too close to ω_n. If at all possible the cutoff point should be five to ten times ω_n. How far ω_c can be placed from ω_n depends on the input frequency relationship to ω_n since f_{in} is, after all, what is being filtered. A side benefit of this simple RC pulse "flattener" is a reduction in f_{in} sidebands around f_{out} for synthesizers with $N > 1$. However, a series of RC filters is not recommended for either extended

Courtesy Motorola Semiconductor Products, Inc.

225

pulse suppression or sideband improvement as excess phase will begin to build up at the loop crossover ($\approx \omega_n$) and tend to cause instability. This will be discussed in more detail later.

Spurious Outputs

Although the major problem in phase-locked loop design is defining loop gain and phase margin under dynamic operating conditions, high-quality synthesizer designs also require special consideration to minimize spurious spectral components — the worst of which is reference-frequency sidebands. Requirements for good sideband suppression often conflict with other performance goals — loop dynamic behavior, suppression of VCO noise, or suppression of other in-loop noise. As a result, most synthesizer designs require compromised specifications. For a given set of components and loop dynamic conditions, reference sidebands should be predicted and checked against design specifications before any hardware is built.

Any steady-state signal on the VCO control will produce sidebands in accordance with normal FM theory. For small spurious deviations on the VCO, relative sideband-to-carrier levels can be predicted by:

$$\frac{\text{sidebands}}{\text{carrier}} \cong \frac{V_{ref} K_V}{2\omega_{ref}} \quad (13)$$

where V_{ref} = peak voltage value of spurious frequency at the VCO input.

Unwanted control line modulation can come from a variety of sources, but the most likely cause is phase detector pulse components feeding through the loop filter. Although the filter does establish loop dynamic conditions, it leaves something to be desired as a low pass section for reference frequency components.

For the usual case where ω_{ref} is higher than $1/T_2$, the K_F function amounts to a simple resistor ratio:

$$K_F(j\omega)\bigg|_{\omega=\omega_{ref}} \cong -\frac{R_2}{R_1} \quad (14)$$

By substitution of Equations 9 and 10, this signal transfer can be related to loop parameters.

$$K_F(j\omega)\bigg|_{\omega=\omega_{ref}} \cong \frac{2\zeta N\omega_n}{K_\phi K_V} = \frac{V_{ref}}{V_\phi} \quad (15)$$

where V_{ref} = peak value of reference voltage at the VCO input, and
V_ϕ = peak value of reference frequency voltage at the phase detector output.

Sideband levels relative to reference voltage at the phase detector output can be computed by combining Equations 13 and 15:

$$\frac{\text{sideband level}}{f_{out} \text{ level}} = V_\phi \left(\frac{\zeta N \omega_n}{\omega_{ref} K_\phi} \right) \quad (16)$$

From Equation 16 we find that for a given phase detector, a given value of R_1 (which determines V_ϕ), and given basic system constraints (N, f_{ref}), only ζ and ω_n remain as variables to diminish the sidebands. If there are few limits on ω_n, it may be lowered indefinitely until the desired degree of suppression is obtained. If ω_n is not arbitrary and the sidebands are still objectionable, additional filtering is indicated.

One item worthy of note is the absence of K_V in Equation 16. From Equation 15 it might be concluded that decreasing K_V would be another means for reducing spurious sidebands, but for constant values of ζ and ω_n this is not a free variable. In a given loop, varying K_V will certainly affect sideband voltage, but will also vary ζ and ω_n.

On the other hand, the choice of ω_n may well affect spectral purity near the carrier, although reference sideband levels may be quite acceptable.

In computing sideband levels, the value of V_ϕ must be determined in relation to other loop components. Residual reference frequency components at the phase detector output are related to the dc error voltage necessary to supply charge pump leakage current and amplifier bias current. From these average voltage figures, spectral components of the reference frequency and its harmonics can be computed using an approximation that the phase detector output consists of square waves τ seconds wide repeated at t second intervals (Figure 17). A Fourier anal-

FIGURE 17 — PHASE DETECTOR OUTPUT

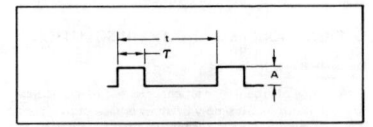

ysis can be summarized for small ratios of τ/t by:

(1) the average voltage (V_{avg}) is $A(\tau/t)$
(2) the peak reference voltage value (V_ϕ) is twice V_{avg}, and
(3) the second harmonic ($2f_{ref}$) is roughly equal in amplitude to the fundamental.

By knowing the requirements for (1) due to amplifier bias and leakage currents, values for (2) and (3) are uniquely determined.

An example of this sideband approximation technique can be illustrated using the parameters specified for the synthesizer design included in the applications information section.

$N_{max} = 30$ $\omega_n = 4500$
$K_V = 11.2 \times 10^6$ rad/s/V $R_1 = 2$ kΩ
$K_\phi = 0.12$ V/rad $f_{ref} = 100$ kHz
$\zeta = 0.8$

Substituting these numbers into Equation 16:

Courtesy Motorola Semiconductor Products, Inc.

$$\frac{\text{sideband}}{f_{out}} = V_\phi \frac{(0.8)(30)(4500)}{2\pi(10^5)(0.111)} \quad (17)$$

$$= V_\phi (1.55) \quad (18)$$

The result illustrates how much reference feedthrough will affect sideband levels. If 1.0 mV peak of reference appears at the output of the phase detector, the nearest sideband will be down 56.2 dB.

If the amplifier section included in the MC4344/4044 is used, with $R_L = 1\ k\Omega$, some approximations of the value of V_ϕ can be made based on the input bias current and the value of R_1. The phase detector must provide sufficient average voltage to supply the amplifier bias current, I_b, through R_1; when the bias current is about 5.0 μA and R_1 is 2 kΩ, V_{avg} must be 10 mV. From the assumptions earlier concerning the Fourier transform, and with the help of Figure 18, we can see that the phase detector duty cycle will be about 1.7% (A = 0.6 V), giving

FIGURE 18 — OUTPUT ERROR CHARACTERISTICS

DUTY CYCLE (%)	PHASE ERROR (Deg)	V_{avg} (mV)	V_ϕ(peak) (mV)
0.1	0.36	0.6	1.2
0.2	0.72	1.2	2.4
0.3	1.08	1.8	3.6
0.4	1.44	2.4	4.8
0.5	1.80	3.0	6.0
0.6	2.16	3.6	7.2
0.7	2.52	4.2	8.4
0.8	2.88	4.8	9.6
0.9	3.24	5.4	10.8
1.0	3.60	6.0	12.0
2.0	7.2	12.0	24.0
3.0	10.8	18.0	35.9
4.0	14.4	24.0	47.9
5.0	18.0	30.0	59.8
6.0	21.6	36.0	71.6
7.0	25.2	42.0	83.3
8.0	28.8	48.0	95.0
9.0	32.4	54.0	106.6
10.0	36.0	60.0	118.0

a fundamental (reference) of 20 mV peak. If this value for V_ϕ is substituted into Equation 18, the resulting sideband ratio represents 30 dB suppression due to this component alone.

For loop amplifiers having very high gains and relatively low bias currents, another factor to consider is reverse leakage current, I_L, of the MC4344/4044 charge pump. This is generally less than 1.0 μA although it is no more than 5.0 μA over the temperature range. A typical value for design for room temperature operation is 0.1 μA. To minimize the effects of amplifier bias and leakage currents a relatively small value of R_1 should be chosen. A minimum value of 1 kΩ is a good choice.

After values for C and R_2 have been computed on the basis of loop dynamic properties, the overall sideband to f_{out} ratio computation can be simplified.

Since
$$V_\phi = 2 V_{avg}$$
$$V_{avg} = (I_b + I_L) R_1$$
$$V_\phi = 2(I_b + I_L) R_1$$

$$V_{ref} = V_\phi \left(\frac{R_2}{R_1}\right)$$
$$= 2R_1(I_b + I_L)\left(\frac{R_2}{R_1}\right) = 2R_2(I_b + I_L)$$

we find that

$$\frac{\text{sideband}}{f_{out}} = \frac{V_{ref} K_V}{2\omega_{ref}} \quad (19)$$

$$\frac{\text{sideband}}{f_{out}} = \frac{2R_2(I_b + I_L)K_V}{2\omega_{ref}} \quad (20)$$

Equation 20 indicates that excellent suppression could be achieved if the bias and leakage terms were nulled by current summing at the amplifier input (Figure 19). This

FIGURE 19 — COMPENSATING FOR BIAS AND LEAKAGE CURRENT

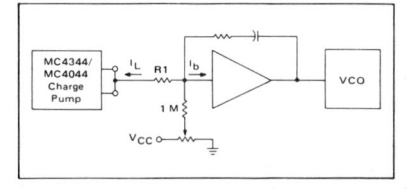

has indeed proved to be the case. Experimental results indicate that greater than 60 dB rejection can routinely be achieved at a constant temperature. However when nulling fairly large values (> 100 nA), the rejection becomes quite sensitive since leakages are inherently a function of temperature. This technique has proved useful in achieving improved system performance when used in conjunction with good circuit practice and reference filtering.

Additional Loop Filtering

So far, only the effects of fundamental loop dynamics on resultant sidebands have been considered. If further sideband suppression is required, additional loop filtering is indicated. However, care must be taken in placement of any low pass rolloff with regard to the loop natural frequency (ω_n). On one hand, the "corner" should be well below (lower than) ω_{ref} and yet far removed (above) from ω_n. Although no easy method for placing the rolloff point exists, a rule of thumb that usually works is:

$$\omega_c = 5\omega_n \quad (21)$$

Courtesy Motorola Semiconductor Products, Inc.

Reference frequency suppression per pole is the ratio of ω_c to ω_{ref}.

$$SB_{dB} \cong n\ 20\ \log_{10}\left(\frac{\omega_c}{\omega_{ref}}\right) \qquad (22)$$

where n is the number of poles in the filter.

Equation 22 gives the additional loop suppression to ω_{ref}; this number should be added to whatever suppression already exists.

For non-critical applications, simple RC networks may suffice, but if more than one section is required, loop dynamics undergo undesirable changes. Loop damping factor decreases, resulting in a high percentage of overshoot and increased ringing since passive RC sections tend to accumulate phase shift more rapidly than signal suppression and part of this excess phase subtracts from the loop phase margin. Less phase margin translates into a lower damping factor and can, in the limit, cause outright oscillation.

A suitable alternative is an active RC section, Figure 20,

FIGURE 20 — OPERATIONAL AMPLIFIER LOW PASS FILTER

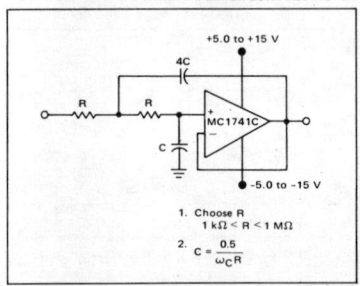

1. Choose R
   $1\ k\Omega < R < 1\ M\Omega$
2. $C = \dfrac{0.5}{\omega_c R}$

compatible with the existing levels and voltages. An active two pole filter (second order section) can realize a more gradual phase shift at frequencies less than the cutoff point and still get nearly equal suppression at frequencies above the cutoff point. Sections designed with a slight amount of peaking ($\zeta \cong 0.5$) show a good compromise between excess phase below cutoff (ω_c), without peaking enough to cause any danger of raising the loop gain for frequencies above ω_n. A fairly non-critical section may simply use an emitter follower as the active device with two resistors and capacitors completing the circuit (Figure 21). This provides a -12 dB/octave (-40 dB/decade) rolloff characteristic above ω_n, though the attenuation may be more accurately determined by Equation 22. If the sideband problem persists, an additional section may be added in series with the first. No more than two sections are recommended since at that time either (1) the constraint

FIGURE 21 — EMITTER FOLLOWER LOW PASS FILTER

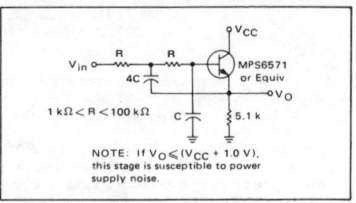

NOTE: If $V_O \leqslant (V_{CC} + 1.0\ V)$, this stage is susceptible to power supply noise.

between ω_n and ω_{ref} is too close, or (2) reference voltage is modulating the VCO from a source other than the phase detector through the loop amplifier.

VCO Noise

Effects of noise within the VCO itself can be evaluated by considering a closed loop situation with an external noise source, e_n, introduced at the VCO (Figure 22). Re-

FIGURE 22 — EFFECTS OF VCO NOISE

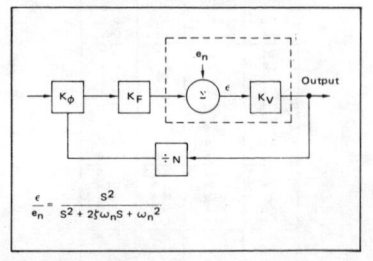

$$\frac{\epsilon}{e_n} = \frac{S^2}{S^2 + 2\zeta\omega_n S + \omega_n^2}$$

sultant modulation of the VCO by error voltage, ϵ, is a second order high pass function:

$$\frac{\epsilon}{e_n} = \frac{S^2}{S^2 + \dfrac{ST_2 K_\phi K_V}{T_1 N} + \dfrac{K_\phi K_V}{T_1 N}} \qquad (23)$$

$$= \frac{S^2}{S^2 + 2\zeta\omega_n + \omega_n^2}$$

This function has a slope of 12 dB/octave at frequencies less than ω_n (loop natural frequency), as shown in Figure 23. This means that noise components in the VCO above ω_n will pass unattenuated and those below will have some degree of suppression. Therefore choice of loop natural frequency may well rest on VCO noise quality.

Other Spurious Responses

Spurious components appearing in the output spectrum are seldom due to reference frequency feedthrough alone.

Courtesy Motorola Semiconductor Products, Inc.

FIGURE 23 – LOOP RESPONSE TO VCO NOISE

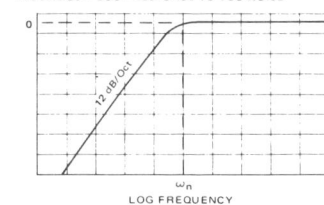

across the load (unless remote sensing is used). One solution to the ground-coupled noise problem is to lay out the return path with the most sensitive regulated circuit at the farthest point from power supply entry as shown in Figure 25.

FIGURE 25 – REGULATOR LAYOUT

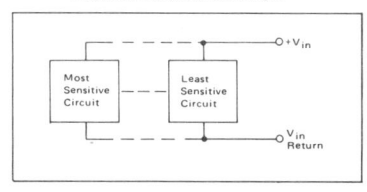

Modulation of any kind appearing on the VCO control line will cause spurious sidebands and can come in through the loop amplifier supply, bias circuitry in the control path, a translator, or even the VCO supply itself. Some VCO's have a relatively high sensitivity to power supply variation. This should be investigated and its effects considered. Problems of this nature can be minimized by operating all devices except the phase detector, charge pump, and VCO from a separate and well isolated supply. A common method uses a master supply of about 10 or 12 volts and two regulators to produce voltages for the PLL – one for all the logic (including the phase detector) and the other for all circuitry associated with the VCO control line.

Sideband and noise performance is also a function of good power supply and regulator layout. As mentioned earlier, extreme care should be exercised in isolating the control line voltage to the VCO from influences other than the phase detector. This not only means good voltage regulation but ac bypassing and adherence to good grounding techniques as well. Figure 24 shows two separate

FIGURE 24 – LOOP VOLTAGE REGULATION

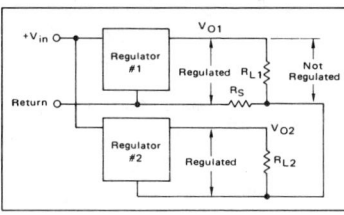

regulators and their respective loads. Resistor R_S is a small stray resistance due to a common thin ground return for both R_{L1} and R_{L2}. Any noise in R_{L2} is now reproduced (in a suppressed form) across R_{L1}. Load current from R_{L1} does not affect the voltage across R_{L2}. Even though the regulators may be quite good, they can hold V_O constant only across their outputs, not necessarily

Even for regulated subcircuits, accumulated noise on the ground bus can pose major problems since although the cross currents do not produce a differential load voltage directly, they do produce essentially common mode noise on the regulators. Output differential load noise then is a function of the input regulation specification. By far the best way to sidestep the problem is to connect each subcircuit ground to the power supply entry return line as shown in Figure 26.

FIGURE 26 – REGULATOR GROUND CONNECTION

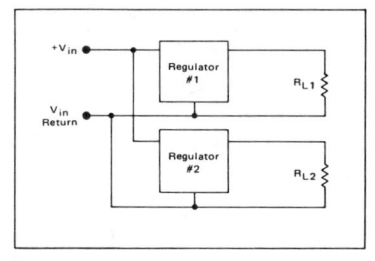

In Figures 24 and 26, R_{L1} and R_{L2} represent component groups in the system. The designer must insure that all ground return leads in a specific component group are returned to the common ground. Probably the most overlooked components are bypass capacitors. To minimize sidebands, extreme caution must be taken in the area immediately following the phase detector and through the VCO. A partial schematic of a typical loop amplifier and filter is shown in Figure 27 to illustrate the common grounding technique.

Courtesy Motorola Semiconductor Products, Inc.

FIGURE 27 — PARTIAL SCHEMATIC OF LOOP AMPLIFIER AND FILTER

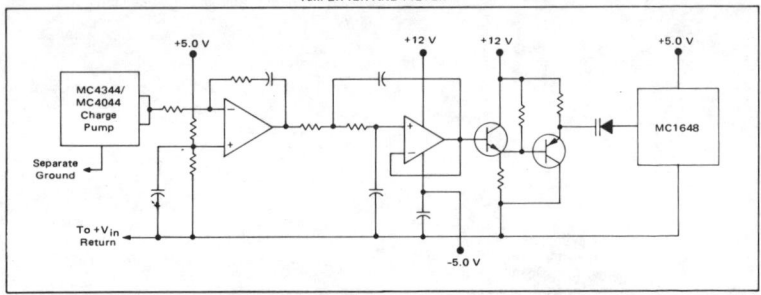

Bypassing in a phase-locked loop must be effective at both high frequencies and low frequencies. One capacitor in the 1.0-to-10 μF range and another between 0.01 and 0.001 μF are usually adequate. These can be effectively utilized both at the immediate circuitry (between supply and common ground) and the regulator if it is some distance away. When used at the regulator, a single electrolytic capacitor on the output and a capacitor pair at the input is most effective (Figure 28). It is important, again, to note that these bypasses go from the input/output pins to as near the regulator ground pin as possible.

FIGURE 28 — SUGGESTED BYPASSING PROCEDURE

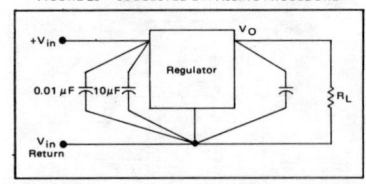

APPLICATIONS INFORMATION

Frequency Synthesizers

The basic PLL discussed earlier is actually a special case of frequency synthesis. In that instance, $f_{out} = f_{in}$, although normally a programmable counter in the feedback loop insures the general rule that $f_{out} = Nf_{in}$ (Figure 29). In the synthesizer f_{in} is usually constant (crystal controlled) and f_{out} is changed by varying the programmable divider (÷ N). By stepping N in integer increments, the output frequency is changed by f_{in} per increment. In communication use, this input frequency is called the "channel spacing" or, in general, it is the reference frequency.

There is essentially no difference in loop dynamic problems between the basic PLL and synthesizers except that synthesizer designers must contend with problems peculiar to loops where N is variable and greater than 1. Also, sidebands or spectral purity usually require special attention. These and other aspects are discussed in greater detail in AN-535. The steps for a suitable synthesis procedure may be summarized as follows:

FIGURE 29 — PHASE-LOCKED LOOP WITH PROGRAMMABLE COUNTER

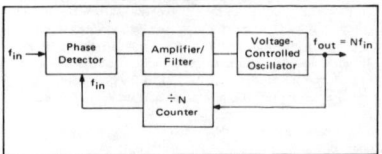

Synthesis Procedure

1. Choose input frequency. (f_{ref} = channel spacing)
2. Compute the range of digital division:

$$N_{max} = \frac{f_{max}}{f_{ref}}$$

$$N_{min} = \frac{f_{min}}{f_{ref}}$$

3. Compute needed VCO range:

$$(2f_{max} - f_{min}) < f_{VCO} < (2f_{min} - f_{max})$$

4. Choose minimum ζ from transient response plot, Figure 9. A good starting point is $\zeta = 0.5$.

Courtesy Motorola Semiconductor Products, Inc.

5. Choose ω_n from needed response time (Figure 9):

$$\omega_n = \frac{\omega_n t}{t}$$

6. Compute C:

$$C = \frac{K_\phi K_V}{N_{max}\omega_n^2 R_1}$$

7. Compute R_2:

$$R_2 = \frac{2\zeta_{min}}{\omega_n C}$$

8. Compute ζ_{max}:

$$\zeta_{max} = \zeta_{min}\sqrt{\frac{N_{max}}{N_{min}}}$$

9. Check transient response of ζ_{max} for compatibility with transient specification.
10. Compute expected sidebands:

$$\frac{sideband}{f_{out}} \cong \frac{(I_b + I_L)R_2 K_V}{\omega_{ref}} \quad (A)$$

(I_L is about 100 nA at $T_J = 25°C$.)

11. If step 10 yields larger sidebands than are acceptable, add a single pole at the loop amplifier by splitting R_1 and adding C_c as shown in Figure 15:

$$C_c \cong \frac{0.8}{R_1 \omega_n}$$

Added sideband suppression (dB) is:

$$dB \cong 20 \log_{10} \frac{1}{\sqrt{1 + \frac{\omega_{ref}^2}{25(\omega_n)^2}}} \quad (B)$$

12. If step 11 still does not give the desired results, add a second order section at $\omega_c = 5\omega_n$ using either the configuration of Figure 20 or 21. The expected improvement is twice that of the single pole in step 10.

$$dB \cong 40 \log_{10} \frac{1}{\sqrt{1 + \frac{\omega_{ref}^2}{25(\omega_n)^2}}} \quad (C)$$

Total sideband rejection is then the total of 20 log$_{10}$(A) + (B) + (C).

Design Example (Figure 30)

Assume the following requirements:
Output frequency, f_{out} = 2.0 MHz to 3.0 MHz
Frequency steps, f_{in} = 100 kHz

FIGURE 30 – CIRCUIT DIAGRAM OF TYPE 2 PHASE-LOCKED LOOP

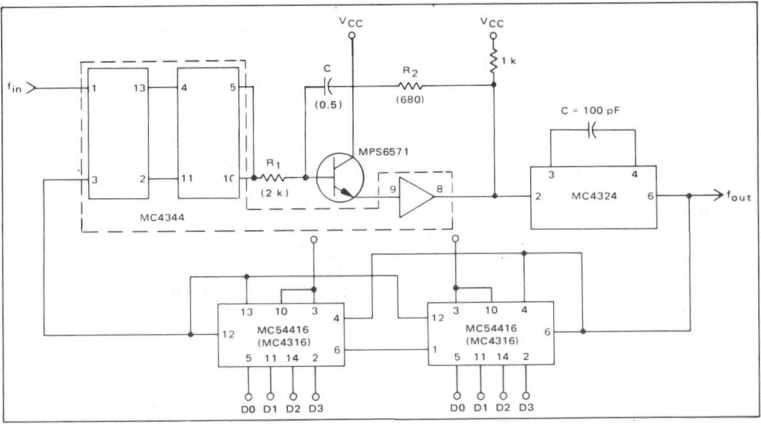

Courtesy Motorola Semiconductor Products, Inc.

Lockup time between channels (to 5%) = 1.0 ms
Overshoot < 20%.
Minimum sideband suppression = -30 dB

From the steps of the synthesis procedure:

1. $f_{ref} = f_{in}$ = 100 kHz

2. $N_{max} = \dfrac{f_{max}}{f_{ref}} = \dfrac{3.0 \text{ MHz}}{0.1 \text{ MHz}} = 30$

   $N_{min} = \dfrac{f_{min}}{f_{ref}} = \dfrac{2.0 \text{ MHz}}{0.1 \text{ MHz}} = 20$

3. VCO range:
   The VCO output frequency range should extend beyond the specified minimum-maximum limits to accommodate the overshoot specification. In this instance f_{out} should be able to cover an additional 20% on either end. End limits on the VCO are:

   $f_{out}max \geqslant 3.0 + 0.2(3.0) = 3.6$ MHz
   $f_{out}min \leqslant 2.0 - 0.2(2.0) = 1.6$ MHz

   This VCO range ($\approx$ 2.25:1) is realizable with the MC4324/4024 voltage controlled multivibrator. From Figure 7 of the MC4324/4024 data sheet we find the required tuning capacitor value to be 120 pF and the VCO gain, K_V, typically 11 x 10^6 rad/s/v.

4. From the step response curve of Figure 5, ζ = 0.8 will produce a peak overshoot less than 20%.

5. Referring to Figure 9, overshoot with ζ = 0.8 will settle to within 5% at $\omega_n t$ = 4.5. Since the required lock-up time is 1.0 ms,

   $\omega_n = \dfrac{\omega_n t}{t} = \dfrac{4.5}{t} = \dfrac{4.5}{0.001} = (4.5)(10^3)$ rad/s

6. In order to compute C, phase detector gain and R1 must be selected. Phase detector gain, K_ϕ, for the MC4344/4044 is approximately 0.1 volt/radian with R_1 = 1 kΩ. Therefore,

   $C = \dfrac{(0.1)(11 \times 10^6)}{(30)(4.5 \times 10^3)^2(10^3)} = 2.0 \mu F$

7. At this point, R_2 can be computed:

   $R_2 = \dfrac{2\zeta_{min}}{\omega_n C} = \dfrac{1.6}{(4.5 \times 10^3)(2 \times 10^{-6})} = 180 \Omega$

8. $\zeta_{max} = \zeta_{min} \sqrt{\dfrac{N_{max}}{N_{min}}} = 0.98$

9. Figure 9 shows that ζ = 0.98 will meet the settling time requirement.

10. Sidebands may be computed for two cases: (1) with I_L (charge pump leakage current) nominal (100 nA), and (2) with I_L maximum (5.0 μA).

$\left.\dfrac{\text{sideband}}{f_{out}}\right|_{max} = \dfrac{(5 \times 10^{-6})(180)(11 \times 10^6)}{4.5 \times 10^3} \cong 2.2$

Since I_L (nominal) is 50 times lower than I_L (maximum), the sideband-to-center frequency ratio nominally would be:

$\left.\dfrac{\text{sideband}}{f_{out}}\right|_{nom} = \dfrac{2.2}{50} = 0.044$

$= 20 \log_{10}(0.044) \cong -27$ dB

This suppression figure does not meet the original design requirement. Therefore further improvements will be made.

11. By splitting R_1 and C_c, further attenuation can be gained. The magnitude of C_c is approximately:

$C_c \cong \dfrac{0.8}{R_1 \omega_n} = \dfrac{0.8}{(10^3)(4.5)(10^3)} \cong 0.2 \mu F$

Improvement in sidebands will be:

$20 \log_{10} \dfrac{1}{1 + \dfrac{105^2}{25(4.5 \times 10^3)^2}} = -13$ dB

Nominal suppression is now -40 dB. Worst-case is 34 dB higher than nominal suppression (50:1 ratio), or -6.0 dB. Therefore additional filtering is required.

12. Additional filters such as second order sections are exactly double the single order sections as designed in step 11. Adding such a filter would give an additional -26 dB rejection factor. Therefore, one second order filter section would result in an overall sideband suppression of -67 dB nominal and -32 dB maximum.

Design of the passive components for the added section with R assigned a value of 10 kΩ is:

$C = \dfrac{0.1}{\omega_n R} = \dfrac{0.1}{(4.5 \times 10^{-3})(10^4)} = 0.2 \mu F$

See Figures 20 and 21 for two configurations that will satisfy this filter requirement.

Clock Recovery from Phase-Encoded Data

The electro-mechanical system used for recording digital data on magnetic tape often introduces random variations in tape speed and data spacing. Because of this and the encoding technqiue used, it is usually necessary to regenerate a synchronized clock from the data during this read cycle. One method for doing this is to phase-lock a voltage controlled multivibrator to the data as it is read (Figure 31).

Courtesy Motorola Semiconductor Products, Inc.

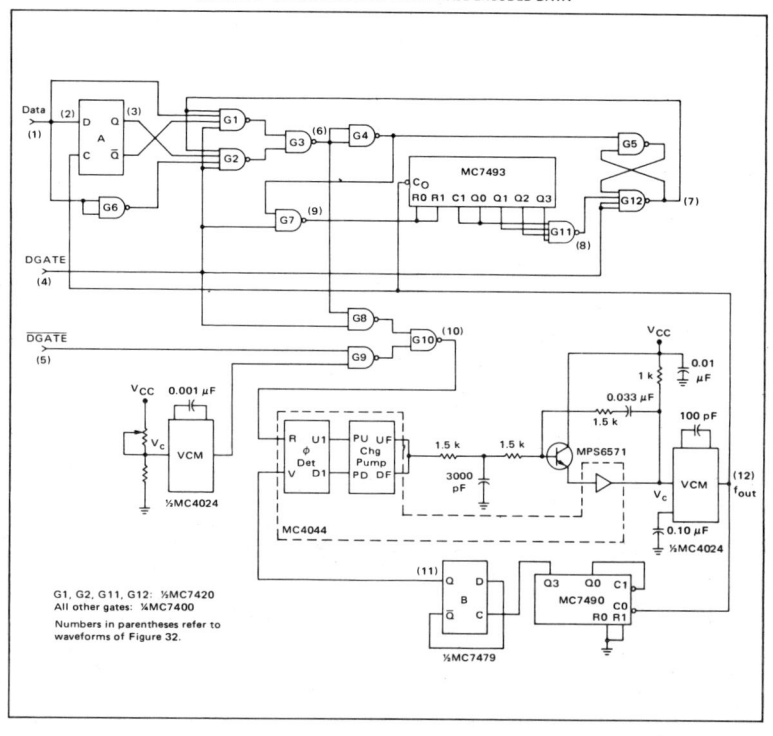

FIGURE 31 – CLOCK RECOVERY FROM PHASE-ENCODED DATA

Courtesy Motorola Semiconductor Products, Inc.

FIGURE 32 – TIMING DIAGRAM – CLOCK RECOVERY FROM
PHASE-ENCODED DATA

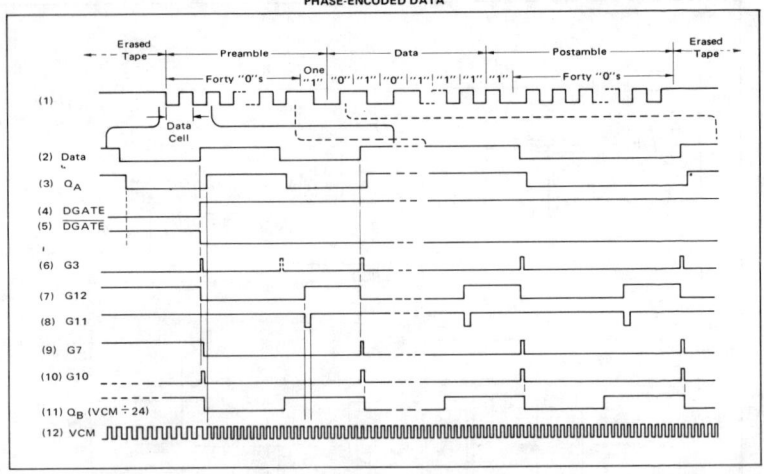

A typical data block using the phase encoded format is shown in row 1 of Figure 32. The standard format calls for recording a preamble of forty "0"s followed by a single "1"; this is followed by from 18 to 2048 characters of data and a postamble consisting of a "1" followed by forty "0"s. The encoding format records a "0" as a transition from low to high in the middle of a data cell. A "1" is indicated by a transition from high to low at the data cell midpoint. When required, phase transitions occur at the end of data cells. If a string of either consecutive "0"s or consecutive "1"s is recorded, the format duplicates the original clock; the clock is easily recovered by straight forward synchronization with a phase-locked loop. In the general case, where the data may appear in any order, the phase-encoded data must be processed to obtain a single pulse during each data cell before it is applied to the phase detector. For example, if the data consisted only of alternating "1"s and "0"s, the phase-encoded format would result in a waveform equal to one-half the original clock frequency. If this were applied directly to the loop, the VCM would of course move down to that frequency. The encoding format insures that there will be a transition in the middle of each data time. If only these transitions are sensed they can be used to regenerate the clock. The schematic diagram of Figure 31 indicates one method of accomplishing this.

The logic circuitry generates a pulse at the midpoint of each data cell which is then applied to the reference input of the phase detector. The loop VCM is designed to operate at some multiple of the basic clock rate. The VCM frequency selected depends on the decoding resolution desired and other system timing requirements. In this example, the VCM operates at twenty-four times the clock rate (Figure 32, Row 12).

Referring to Figure 31 and the timing diagram of Figure 32, the phase-encoded data (Figure 32, Row 1) is combined with a delayed version of itself (output of flip-flop A row 3) to provide a positive pulse out of G3 for every transition of the input signal. Portions of the data block are shown expanded in row 2 of Figure 32. Flip-flop A delays the incoming data of one-half of a VCM clock period. Gates G1, G2, and G3 implement the logic Exclusive OR of waveforms 1 and 3 except when inhibited by DGATE (row 4) or the output of G12 (row 7). DGATE and its complement, $\overline{\text{DGATE}}$, serve to initialize the circuitry and insure that the first transition of the data block (a phase transition) is ignored. The MC7493 binary counter and the G5-G12 latch generate a suitable signal for gating out G3 pulses caused by phase transitions at the end of a data cell, such as the one shown dashed in row 6.

The initial data pulse from G3 sets G12 low and is combined with DGATE in G7 to reset the counter to its zero state. Subsequent VCM clock pulses now cycle the counter and approximately one-third of the way through the next data cell the counter's full state is decoded by G11, generating a negative transition. This causes G12 to go high, removing the inhibit signal until it is again reset by the next data transition. This pulse also resets the

Courtesy Motorola Semiconductor Products, Inc.

234

counter, continuing the cycle and generating a positive pulse at the midpoint of each data cell as required.

Acquisition time is reduced if the loop is locked to a frequency approximately the same as the expected data rate during inter-block gaps. In Figure 31, this is achieved by operating the remaining half of the dual VCM at slightly less than the data rate and applying it to the reference input of the phase detector via the G8-G9-G10 data selector. When data appears, DGATE and $\overline{\text{DGATE}}$ cause the output of G3 to be selected as the reference input to the loop.

The loop parameters are selected as a compromise between fast acquisition and jitter-free tracking once synchronization is achieved. The resulting filter component values indicated in Figure 31 are suitable for recovering the clock from data recorded at a 120 kHz rate, such as would result in a tape system operating at 75 i.p.s. with a recording density of 1600 b.p.i. Synchronization is achieved by approximately the twenty-fourth bit time of the preamble. The relationship between system requirements and the design procedure is illustrated by the following sample calculation:

Assume a -3.0 dB loop bandwidth much less than the input data rate ($\approx$ 120 kHz), say 10 kHz. Further, assume a damping factor of $\zeta = 0.707$. From the expression for loop bandwidth as a function of damping factor and undamped natural frequency, ω_n, calculate ω_n as:

$$\omega_{-3\,dB} = \omega_n \left(1 + 2\zeta^2 + \sqrt{2 + 4\zeta^2 + 4\zeta^4}\right)^{1/2} \quad (24)$$

or for $\omega_{-3\,dB} = (2\pi)10^4$ rad/s and $\zeta = 0.707$:

$$\omega_n = \frac{(2\pi)10^4}{2.06} = (3.05)10^4 \text{ rad/s}$$

As a rough check on acquisition time, assume that lockup should occur not later than half-way through a 40-bit preample, or for twenty 8.34 μs data periods.

$$\omega_n t = (3.05)10^4 (20)(8.34)10^{-6} = 5.1 \quad (26)$$

From Figure 9, the output will be within 2 to 3% of its final value for $\omega_n t \approx 5$ and $\zeta = 0.707$. The filter components are calculated by:

and

$$\frac{K_\phi K_V}{R_1 CN} = \omega_n^2 \quad (27)$$

$$\frac{K_\phi K_V R_2}{R_1 N} = 2\zeta\omega_n \quad (28)$$

where $K_\phi = 0.015$ v/rad
$K_V = (18.2)10^6$ rad/s/volt
$N = 24$ = Feedback divider ratio
$\omega_n = (3.05)10^4$ rad/s
$\zeta = 0.707$

$$\frac{K_\phi K_V}{N} = \frac{(0.115)(18.2)10^6}{24} = (8.72)10^4$$

From Equation 27:

$$R_1 C = \frac{K_\phi K_V}{N\omega_n^2} = \frac{(8.72)10^4}{(3.05)^2 10^8} = (9.34)10^{-5}$$

From Equation 28:

$$\frac{R_2}{R_1} = \frac{2\zeta\omega_n N}{K_\phi K_V} = \frac{2(0.707)(3.04)10^4}{(8.72)10^4} = 0.494 \approx 1/2$$

Let $R_1 = 3.0$ kΩ; then $R_2 = 1.5$ kΩ and

$$C = \frac{(9.34)10^{-5}}{(3.0)10^3} = (3.1)10^{-8}$$

or using a close standard value, use $C = 0.0033 \,\mu$F. Now add the additional prefiltering by splitting R_1 and selecting a time constant for the additional section so that it is large with respect to $R_2 C_2$.

$$10(\tfrac{1}{2}R_1)C_s = R_2 C$$

or

$$C_s = \frac{2R_2 C}{10 R_1} = \frac{2(1.5)10^3(3.3)10^{-8}}{10(3.0)10^3} = 3300 \text{ pF}$$

Circuit diagrams utilizing Motorola products are included as a means of illustrating typical semiconductor applications; consequently, complete information sufficient for construction purposes is not necessarily given. The information has been carefully checked and is believed to be entirely reliable. However, no responsibility is assumed for inaccuracies. Furthermore, such information does not convey to the purchaser of the semiconductor devices described any license under the patent rights of Motorola Inc. or others.

Courtesy Motorola Semiconductor Products, Inc.

CASE 607-05

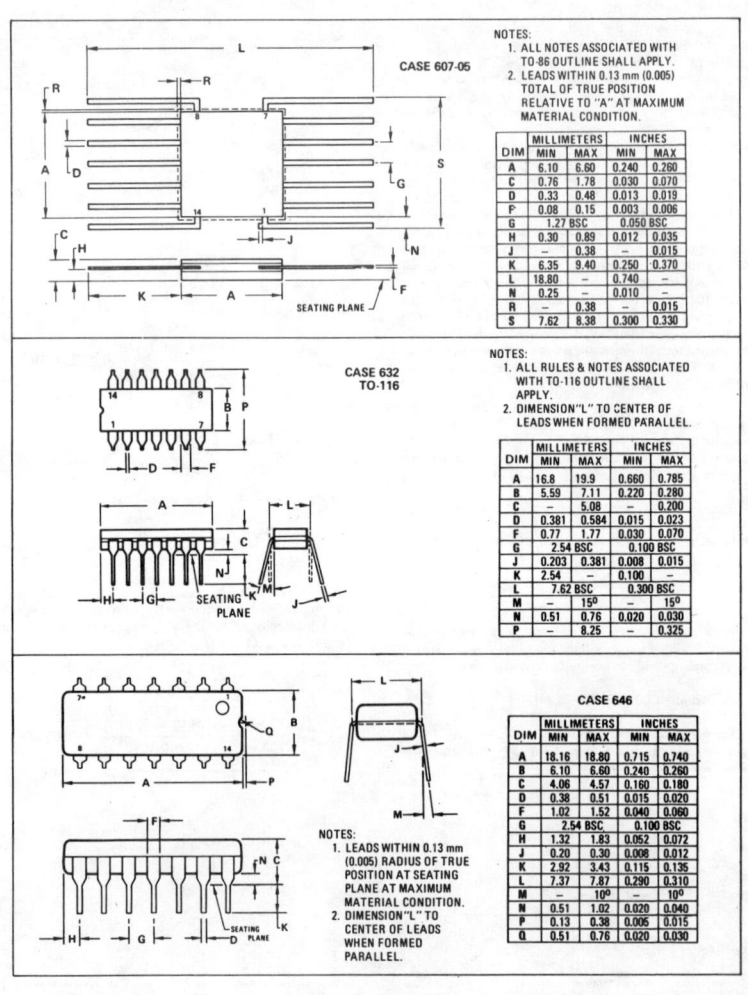

NOTES:
1. ALL NOTES ASSOCIATED WITH TO-86 OUTLINE SHALL APPLY.
2. LEADS WITHIN 0.13 mm (0.005) TOTAL OF TRUE POSITION RELATIVE TO "A" AT MAXIMUM MATERIAL CONDITION.

DIM	MILLIMETERS MIN	MILLIMETERS MAX	INCHES MIN	INCHES MAX
A	6.10	6.60	0.240	0.260
C	0.76	1.78	0.030	0.070
D	0.33	0.48	0.013	0.019
F	0.08	0.15	0.003	0.006
G	1.27 BSC		0.050 BSC	
H	0.30	0.89	0.012	0.035
J	–	0.38	–	0.015
K	6.35	9.40	0.250	0.370
L	18.80	–	0.740	–
N	0.25	–	0.010	–
R	–	0.38	–	0.015
S	7.62	8.38	0.300	0.330

CASE 632 TO-116

NOTES:
1. ALL RULES & NOTES ASSOCIATED WITH TO-116 OUTLINE SHALL APPLY.
2. DIMENSION "L" TO CENTER OF LEADS WHEN FORMED PARALLEL.

DIM	MILLIMETERS MIN	MILLIMETERS MAX	INCHES MIN	INCHES MAX
A	16.8	19.9	0.660	0.785
B	5.59	7.11	0.220	0.280
C	–	5.08	–	0.200
D	0.381	0.584	0.015	0.023
F	0.77	1.77	0.030	0.070
G	2.54 BSC		0.100 BSC	
J	0.203	0.381	0.008	0.015
K	2.54	–	0.100	–
L	7.62 BSC		0.300 BSC	
M	–	15°	–	15°
N	0.51	0.76	0.020	0.030
P	–	8.25	–	0.325

CASE 646

DIM	MILLIMETERS MIN	MILLIMETERS MAX	INCHES MIN	INCHES MAX
A	18.16	18.80	0.715	0.740
B	6.10	6.60	0.240	0.260
C	4.06	4.57	0.160	0.180
D	0.38	0.51	0.015	0.020
F	1.02	1.52	0.040	0.060
G	2.54 BSC		0.100 BSC	
H	1.32	1.83	0.052	0.072
J	0.20	0.30	0.008	0.012
K	2.92	3.43	0.115	0.135
L	7.37	7.87	0.290	0.310
M	–	10°	–	10°
N	0.51	1.02	0.020	0.040
P	0.13	0.38	0.005	0.015
Q	0.51	0.76	0.020	0.030

NOTES:
1. LEADS WITHIN 0.13 mm (0.005) RADIUS OF TRUE POSITION AT SEATING PLANE AT MAXIMUM MATERIAL CONDITION.
2. DIMENSION "L" TO CENTER OF LEADS WHEN FORMED PARALLEL.

Courtesy Motorola Semiconductor Products, Inc.

HCTR0320

DESCRIPTION

The HCTR0320 is a CMOS LSI programmable divide by N counter with a phase/frequency detector for frequency synthesis or phase locked loop (PPL) applications. A minimum PLL system can be made using the HCTR0320, a reference oscillator and divider, low pass filter, and voltage controlled oscillator (VCO). More complex systems may use mixers, frequency multipliers, or a dual modulus prescalar. Most system designs constrain the VCO to oscillate at N times the divided reference oscillator frequency (f_{REF}) so changing N by ΔN changes the VCO frequency by the product (ΔN) • (f_{REF}). Thus multiple VCO frequencies can be generated from only one reference oscillator crystal by varying N. This method results in VCO frequencies which have the same fractional error as the reference crystal oscillator frequency.

FEATURES

- HIGH FREQUENCY OPERATION (10 MHZ)
- LOW POWER CMOS
- ON CHIP PHASE/FREQUENCY DETECTOR
- BCD AND/OR BINARY INPUTS FOR N
- ON CHIP ADDER TO PROVIDE OFFSET
- N PROGRAMMABLE FROM 3 TO 1023
- VCO SIGNAL PRECONDITIONING
- OUTPUT FROM ÷ N COUNTER IS PROVIDED
- POLARITY CONTROL ON VCO CORRECTION SIGNAL

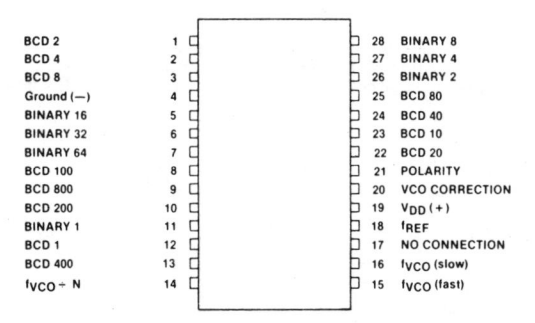

BCD 2	1		28	BINARY 8
BCD 4	2		27	BINARY 4
BCD 8	3		26	BINARY 2
Ground (—)	4		25	BCD 80
BINARY 16	5		24	BCD 40
BINARY 32	6		23	BCD 10
BINARY 64	7		22	BCD 20
BCD 100	8		21	POLARITY
BCD 800	9		20	VCO CORRECTION
BCD 200	10		19	V_{DD} (+)
BINARY 1	11		18	f_{REF}
BCD 1	12		17	NO CONNECTION
BCD 400	13		16	f_{VCO} (slow)
$f_{VCO} \div$ N	14		15	f_{VCO} (fast)

ABSOLUTE MAXIMUM RATINGS	SYM.	VALUE	UNIT
DC Supply Voltage	V_{DD}	+15 to —0.3	Vdc
Input Voltage, All Inputs	V_{in}	V_{DD} to —0.5	Vdc
DC Current Drain Per Pin, All Inputs*	I	10	mAdc
DC Current Drain Per Pin, All Outputs*	I	20	mAdc
Operating Temperature Range	T_A	—40 to 85°C	°C
Storage Temperature Range	T_{stg}	—65 to +150	°C
Power Dissipation	Pd	600 (plastic pkg) 700 (ceramic pkg)	mW

* Protection diodes forward biased

Courtesy Hughes Aircraft Co.

237

EXPLANATION OF BLOCK DIAGRAM

Adder/Decoder This block adds a three digit BCD number (N_{BCD}) to a 7 bit binary number (N_{BIN}) to provide a sum equal to the division integer (N). Each decade of BCD inputs is restricted to valid BCD numbers, zero through nine. The Binary and BCD inputs require full swing signals such as those achieved by SPDT switches or CMOS logic. Positive logic is used.

Programmable Divider - This circuit utilizes a continuously recycling presettable down counter to output a waveform of frequency f_{VCO}/N at a duty cycle of 1/N. f_{VCO} (fast) is the only TTL compatible input and should be used when fast rise and fall times are available and/or maximum speed is required. For input signals with slow rise and fall times such as sine waves, the f_{VCO} (slow) input provides signal preconditioning through a Schmitt Trigger in order to obtain proper rising and falling edges for the digital circuitry. However, the additional circuitry does restrict the maximum operating frequency. The unused f_{VCO} input must be connected to V_{DD} (+). Either f_{VCO} input will accept low frequencies. However, in order to obtain high operating frequencies, dynamic circuitry is used and thus the minimum guaranteed f_{VCO} input frequency is 5 KHz.

Phase/Frequency Detector - This block compares the divider output (f_{VCO}/N) with an external reference frequency (f_{REF}) and generates a correction signal. When the VCO correction output goes from the floating state (NMOS and PMOS switches-off) to V_{DD} (+) or GND (—), the indication is that the leading edges of the two input signals do not occur simultaneously. The leading edge of one signal triggers the correction pulse and the leading edge of the other signal resets the output to the floating state (Refer to Timing Diagram). Therefore, the width of the correction pulse is proportional to the time difference between the leading edges. As the two signals approach equal frequency and phase, the width of the pulse becomes narrower and narrower and the two signals are in "lock". The Polarity input should be tied to V_{DD}(+) if the VCO correction output voltage should decrease to cause an increase in the VCO frequency.

Courtesy Hughes Aircraft Co.

FREQUENCY — f_{VCO} (KHz)

÷ N

f_{REF}

f_{VCO}/N
(VARIABLE FREQ)

"VCO CORRECTION"
(3 STATE OUTPUT)

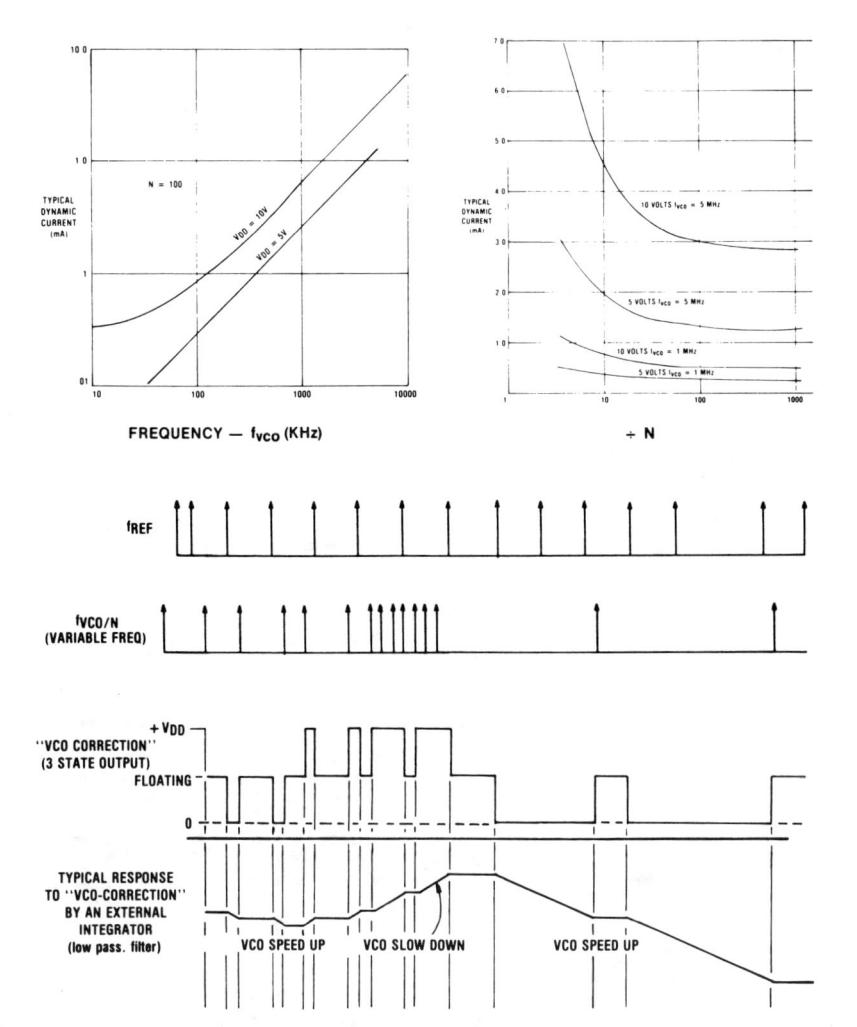

TYPICAL RESPONSE
TO "VCO-CORRECTION"
BY AN EXTERNAL
INTEGRATOR
(low pass. filter)

VCO SPEED UP   VCO SLOW DOWN   VCO SPEED UP

NOTES: 1. ONLY POSITIVE TRANSITIONS OF f_{REF} AND f_{VCO} ARE SHOWN. CIRCUIT OPERATION IS INDEPENDENT OF DUTY CYCLES.
2. POLARITY SENSE IS TIED TO V_{DD}.

TIMING DIAGRAM OF PHASE FREQUENCY DETECTOR

Courtesy Hughes Aircraft Co.

ELECTRICAL SPECIFICATIONS Unless otherwise specified T = −40°C to 85°C V_{DD} tolerance = ±5%

D. C. CHARACTERISTICS		SYMBOL	CONDITIONS	V_{DD}	MIN	MAX	UNITS
Supply Voltage		V_{DD}			4.5	13	V
Input Levels							
BCD and Binary Switches	"1"	V_{IH}		5	4.75	5	V
				10	9.75	10	V
	"0"	V_{IL}		5	0	.25	V
(50 KΩ Impedance required)				10	0	.25	V
f_{VCO} (Fast)	"1"	V_{IH}		5	3.5	5	V
				10	7	10	V
	"0"	V_{IL}		5	0	.4	V
				10	0	1.0	V
f_{VCO} (Slow), f_{REF}	"1"	V_{IH}		5	4.5	5	V
				10	9	10	V
	"0"	V_{IL}		5	0	.5	V
				10	0	1.0	V
Input Leakage Current		I_L	To either V_{DD} or GND	5	—	1	μA
(except BCD and Binary inputs)				10	—	2	μA
Input Capacitance		C_L	(Typical)			5	pf
Output Impedance, f_{VCO}/N and		R_{on}	Within 1 Volt of supply	5	—	500	Ω
VCO Correction				10	—	360	Ω
		R_{off}		5	—		MΩ
A. C. CHARACTERISTICS							
Supply Current		I_{DD}	f_{VCO} = 1 MHz	5	—	.5	mA
Inputs			N = 100	10	—	1.0	mA
f_{VCO} (Fast)							
frequency		F_{VCO}		5	.005	5	MHz
pulse width				10	.010	10	MHz
		PW_H	50% to 50%	5	.10	100	μs
		PW_L		10	.045	50	μs
rise & fall time		t_r, t_f	10% to 90%	5	—	100	ns
				10	—	50	ns
f_{VCO} (Slow)							
frequency		f_{VCO}		5	.005	2.5	MHz
				10	.010	5	MHz
pulse width		PW_H,	50% to 50%	5	.200	100	μs
		PW_L		10	.100	50	μs
rise & fall time		t_r, t_f	10% to 90%	5	No limit		
				10			
f_{REF} pulse width		PW_H,	50% to 50%	5	300	—	ns
		PW_L		10	150	—	ns
rise & fall time		t_r, t_f	10% to 90%	5	—	1	μs
				10	—	1	μs
Outputs							
f_{VCO} (Slow) to f_{VCO}/N propagation		t_{pH}	50% to 50%	5	—	750	ns
delay, falling edge to rising edge			C_L = 10 pf	10	—	420	ns
falling edge to falling edge		t_{pL}	50% to 50%	5	—	680	ns
			C_L = 10 pf	10	—	375	ns
f_{VCO} (Slow) to F_{VCO}/N propagation		t_{pH}	50% to 50%	5	—	360	ns
delay, falling edge to rising edge			C_L = 10 pf	10	—	250	ns
falling edge to falling edge		t_{pL}	50% to 50%	5	—	315	ns
			C_L = 10 pf	10	—	270	ns

Courtesy Hughes Aircraft Co.

CMOS DIGITAL FREQUENCY SYNTHESIZER BLOCK DIAGRAM

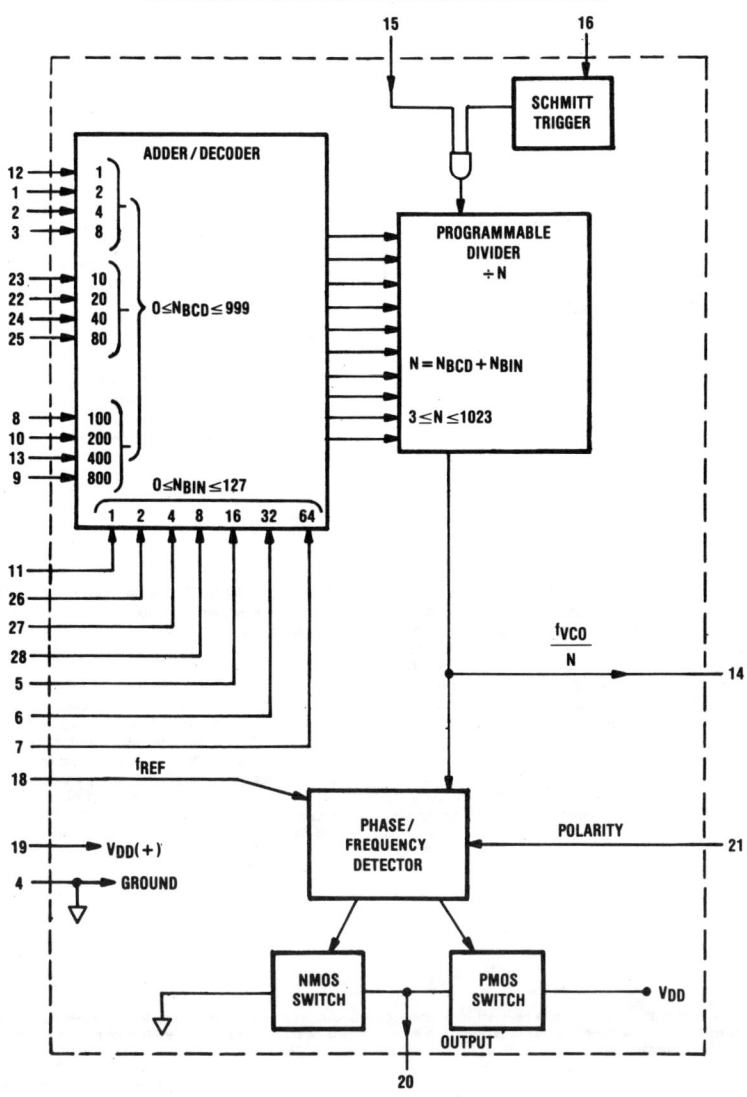

Courtesy Hughes Aircraft Co.

241

TYPICAL DIGITAL FREQUENCY SYNTHESIZER APPLICATION

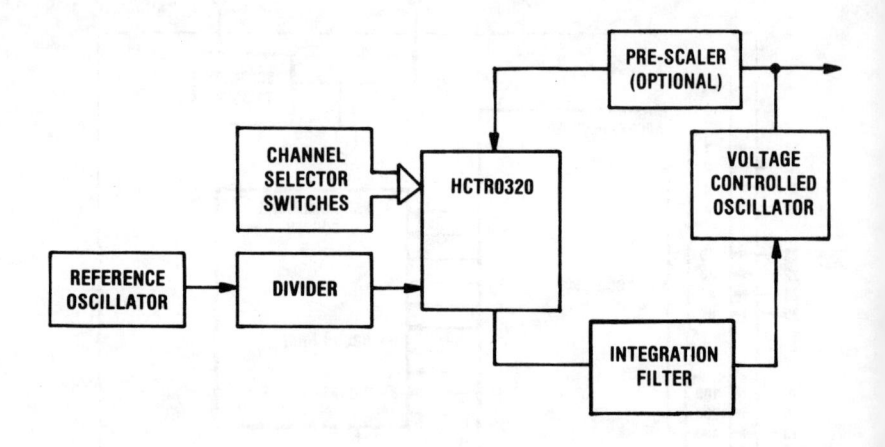

APPLICATION NOTES

The Adder/Decoder, with its BCD and Binary inputs, presents a variety of application opportunities. In some cases it may be desired to input N from three BCD coded thumb wheel switches, in which case the BCD inputs are well suited. If toggle switches are used to set N, then the Binary inputs may be better suited. All unused binary and BCD inputs must be connected to a logic 0 (ground). In some radio transceiver applications it is desirable to offset the transmit and receive frequencies. In these applications, the channel can be set with the BCD inputs and the offset between the transmit and receive frequencies controlled with the Binary inputs (or vice-versa).

Values of 0-999/0-127 can be input on the BCD/Binary lines. However, the maximum N is 1023 and the minimum is 3.

The VCO correction output is a 3 state output which is high, low or floating. When "lock" is achieved, both the NMOS and PMOS output switches are turned off except for very narrow pulses and the output mostly "floats". An integrator and/or low pass filter is required to "smooth out" the pulses and maintain the voltage to the VCO, thus keeping the frequency constant.

Information furnished by Hughes is believed to be accurate and reliable. However, no responsibility is assumed by Hughes for its use; nor for any infringements or patents or other rights of third parties which may result from its use. No license is granted by implication or otherwise under any patent or patent rights of Hughes.

Courtesy Hughes Aircraft Co.

APPENDIX C

Breadboarding Aids

In addition to the SK-10 Universal Breadboarding Socket, E&L Instruments also manufactures an extensive line of useful breadboarding aids, called OUTBOARDS®. This appendix describes the OUTBOARDS that are useful in performing the experiments described in this book, and which were illustrated by schematic diagrams in Chapter 2. Each OUTBOARD attaches directly to the SK-10 socket and obtains the +5 volts and ground power connections from the outer two rows of the solderless terminals. The input and output pins are electrically tied to the sets of 5 solderless terminals in the interior of the breadboarding socket. The following OUTBOARDS are described:

1. LR-2 Logic Switch OUTBOARD
2. LR-4 Seven-Segment LED Display OUTBOARD
3. LR-6 Lamp Monitor OUTBOARD
4. LR-7 Dual Pulser OUTBOARD
5. LR-25 TTL Breadboarding OUTBOARD
6. LR-30 CMOS Breadboarding OUTBOARD
7. LR-31 Function Generator OUTBOARD
8. LR-33 Quartz Crystal OUTBOARD

1. LR-2 Logic Switch OUTBOARD

This OUTBOARD, shown in Fig. C-1, is similar to the schematic shown in Chapter 2 (Fig. 2-6). It provides four logic switches that switch between ground (logic 0) and +5 volts (logic 1), and is used with TTL integrated circuits.

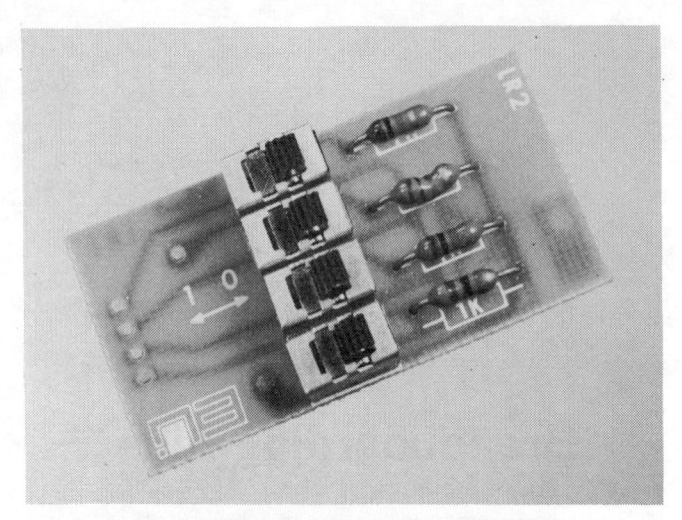

Fig. C-1. The LR-2 logic Switch OUTBOARD.

2. LR-4 Seven-Segment LED Display OUTBOARD

This OUTBOARD, shown in Fig. C-2, is similar to the schematic shown in Chapter 2 (Fig. 2-10). Complete with a decoder/driver and LED display, the LR-4 is a numerical 0 through 9 readout based on the 4-bit bcd input present at its terminals.

Fig. C-2. The LR-4 Seven-Segment LED Display OUTBOARD.

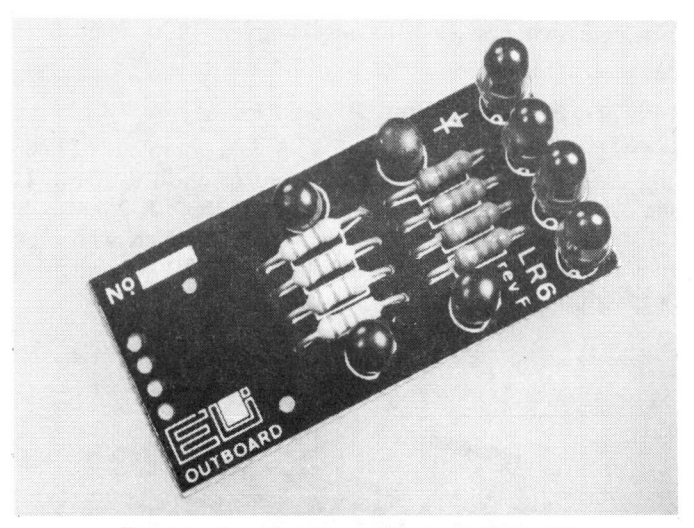

Fig. C-3. The LR-6 Lamp Monitor OUTBOARD.

3. LR-6 Lamp Monitor OUTBOARD

This OUTBOARD, shown in Fig. C-3, is similar to the schematic shown in Chapter 2 (Fig. 2-4). It provides four LEDs which are "off" for a logic 0 input and "on" for a logic 1.

4. LR-7 Dual Pulser OUTBOARD

This OUTBOARD, shown in Fig. C-4, is similar to the schematic shown in Chapter 2 (Fig. 2-8). It contains two independent de-

Fig. C-4. The LR-7 Dual Pulser OUTBOARD.

bounced logic switches, each having complementary logic 0 and 1 outputs.

5. LR-25 TTL Breadboarding OUTBOARD

This OUTBOARD, shown in Fig. C-5, is a complete TTL breadboarding station. It contains the equivalent of one LR-2, two LR-6, and one LR-7 OUTBOARDS. In addition, the LR-25 also has a variable square-wave clock made from the 555 timer, whose output frequency ranges approximately from 0.1 Hz to 20 kHz when an external capacitor (5 pF to 100 μF) is used.

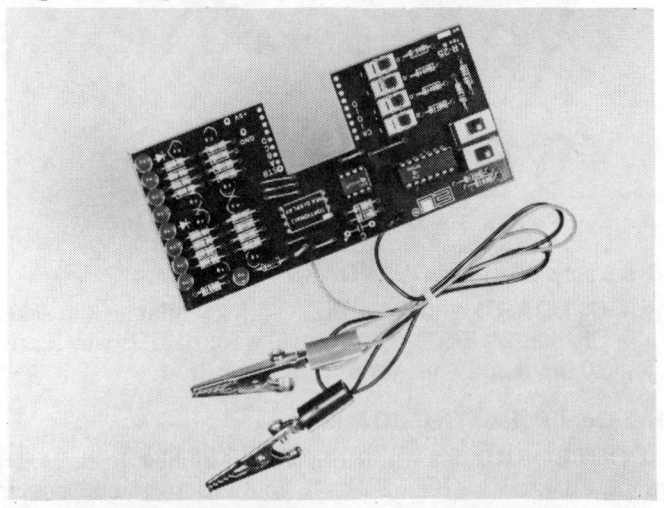

Fig. C-5. The LR-25 TTL Breadboarding OUTBOARD.

6. LR-30 CMOS Breadboarding OUTBOARD

This OUTBOARD, shown in Fig. C-6, is identical in function to the LR-25, but is modified to operate over the range from +3 to +15 volts for CMOS integrated circuits.

7. LR-31 Function Generator OUTBOARD

This OUTBOARD, shown in Fig. C-7, uses an XR-2206 integrated circuit to generate sine, square, and triangle waveforms with adjustable frequency, amplitude, and dc offset. Frequencies of 0.01 Hz to 1 MHz are possible, depending on the value of an external capacitor. With a single capacitor, the dynamic range is greater than 1000:1.

8. LR-33 Quartz Crystal OUTBOARD

This OUTBOARD, shown in Fig. C-8, is similar to the schematic shown in Chapter 2 (Fig. 2-13). With the on-board thumbwheel

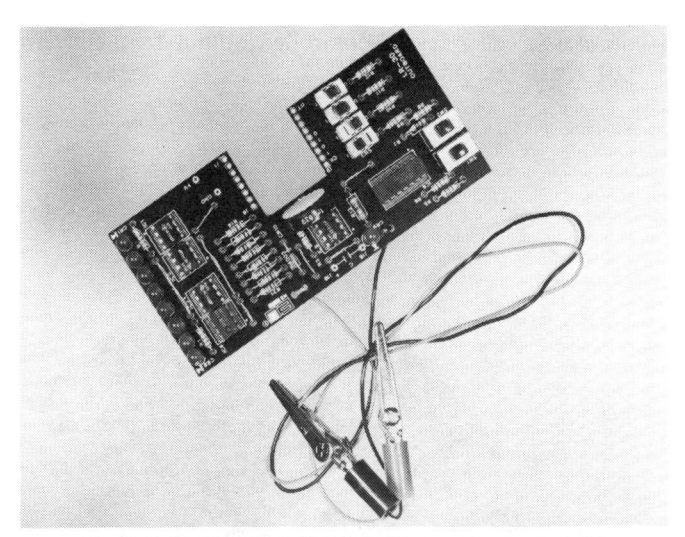

Fig. C-6. The LR-30 CMOS Breadboarding OUTBOARD.

Fig. C-7. The LR-31 Function Generator OUTBOARD.

switch, you can select a crystal-controlled output frequency from 1 MHz to 0.01 Hz in factors of 10.

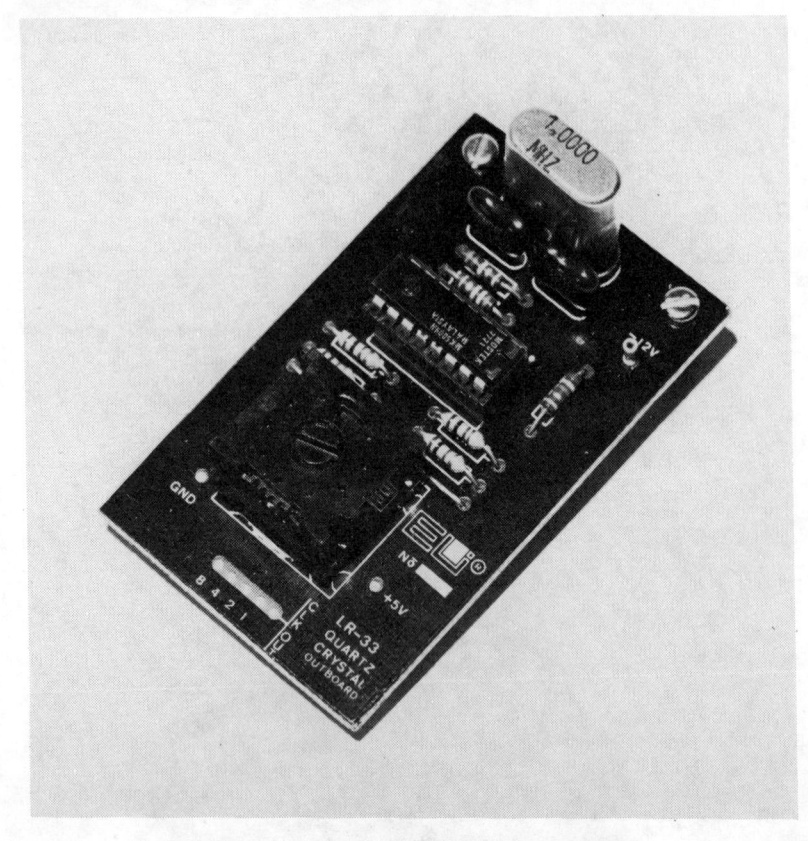

Fig. C-8. The LR-33 Quartz Crystal OUTBOARD.

APPENDIX D

Symbols Used

D = duty cycle (dimensionless)
f_{CH} = synthesizer channel spacing (Hz)
f_H = local oscillator frequency (Hz)
f_i = phase-locked-loop frequency (Hz)
f_{MIX} = mixer output frequency (Hz)
f_{REF} = synthesizer input reference frequency (Hz)
f_o = phase-locked-loop output frequency (Hz)
K = dc loop gain (s^{-1})
K_o = vco conversion gain (rad/s/V)
K_ϕ = phase detector conversion gain (V/rad)
N = modulus of divide-by-N counter
t_s = settling time (s)
T = period of transient oscillations (s)
V_f = vco input voltage, or error voltage (V)
V_o = average (dc) output voltage of phase detector (V)
π = 3.14
ζ = damping factor (dimensionless)
$\Delta\phi$ = input phase difference of phase detector (rad)
ω_C = lock-in range (rad/s)
ω_d = damped natural frequency (rad/s)
ω_i = phase-locked-loop input frequency (rad/s)
ω_L = hold-in range (rad/s)
ω_{LPF} = low-pass-filter cutoff frequency (rad/s)
ω_n = loop natural frequency (rad/s)

Bibliography

The following books and application notes contain a wealth of material on the design, operation, and application of the phase-locked loop. For the most part, however, these are highly technical. On the other hand, the following short articles describe specific applications and are primarily devoid of mathematics.

APPLICATION NOTES

1. Brubaker, R., and Nash, G. *A New Generation of Integrated Avionic Synthesizers.* Motorola AN-553, 1971.
2. Brubaker, R. *An ADF Frequency Synthesizer Utilizing Phase-Lock-Loop I/Cs.* Motorola AN-564, 1972.
3. Connelly, J. A. *A General Analysis of the Phase-Locked Loop.* Harris Semiconductor AN-602, 1972.
4. DeLaune, J. *MTTL and MECL Avionics Digital Frequency Synthesizer.* Motorola AN-532A, 1971.
5. Mills, T. B. *The Phase Locked Loop IC as a Communication System Building Block.* National Semiconductor AN-46, 1971.
6. Nash, G. *Phase-Lock Loop Design Fundamentals.* Motorola AN-535, 1970.
7. Renschler, E., and Welling, B. *An Integrated-Circuit Phase-Locked Loop Digital Frequency Synthesizer.* Motorola AN-463, 1969.

BOOKS

1. Gardner, F. M. *Phaselock Techniques.* John Wiley & Sons, Inc., New York, 1966.
2. Lancaster, D. *CMOS Cookbook.* Howard W. Sams & Co., Inc., Indianapolis, 1977 (Chapter 7).
3. Melen, R., and Garland, H. *Understanding CMOS Integrated Circuits.* Howard W. Sams & Co., Inc., Indianapolis, 1975, pp. 105–109, 117–119.
4. Noll, E. M. *Linear IC Principles, Experiments, and Projects.* Howard W. Sams & Co., Inc., Indianapolis, 1974, pp. 199–213.

SHORT ARTICLES AND APPLICATIONS

1. Allen, G. R. "Synthesize Yourself." *73*, October, 1977, pp. 182–188.
2. Amon, L. E. S., and Lohrey, B. "Versatile Phase Detector Produces Unambiguous Output." *Electronics*, September 15, 1977, pp. 117–119.
3. *Analog Manual*. Signetics Corp., 1976.
4. Bertini, P. J., and VanHooft, R. "A Practical Approach to Two-Meter Frequency Synthesis." *QST;* Part I: June, 1973, pp. 32–36; Part II: July, 1973, pp. 34–39.
5. Cohen, H. "How Phase-Locked Loops Work." *Popular Electronics*, February, 1975, pp. 32–34.
6. ——— "Experimenting With Phase-Locked Loops," *Popular Electronics*, October, 1975, pp. 47–50.
7. Cohen, M. I. "A Practical 2m Synthesizer." *73*, September, 1977, pp. 146–151.
8. Cox, J. "One-Crystal Frequency Synthesizer for Two-Meter FM." *Ham Radio*, September, 1973, pp. 30–38.
9. Delagrange, A. D. "Lock Onto Frequency With Frequency-Lock Loops." *Electronic Design*, June 21, 1977, pp. 84–87.
10. Ferris, R. K. "Constant-Bandwidth PPL Tone Decoder Accepts Wide Range of Input Voltages." *Electronic Design*, November 8, 1977, p. 106.
11. Grebene, A. B. "The Monolithic Phase-Lock-Loop—A Versatile Building Block." *IEEE Spectrum*, March, 1971, pp. 38–49.
12. Hanisko, J. "Timer/Counter Functions as Phase-Locked Loop Component." *EDN*, March 20, 1976, p. 98.
13. Helfrick, A. D. "High-Frequency Frequency Synthesizer." *Ham Radio*, October, 1972, pp. 16–21.
14. Illingworth, G., and Terman, M. "Phase Locked Loop: An Application in Temperature Telemetry and a Method for Its Evaluation." *Physiology and Behavior*, vol. 13, 1974, pp. 335–338.
15. Levy, E. I. "Retriggerable One-Shot Prevents False Triggering of PPL Tone Detector." *Electronic Design*, January 4, 1973, p. 102.
16. Mims, F. M. "The 567 Tone Decoder." *Popular Electronics*, August, 1976, pp. 91–93.
17. More, A. W. "Phase-Locked Loops for Motor-Speed Control." *IEEE Spectrum*, April, 1973, pp. 61–67.
18. Murthi, E. "Monolithic Phase-Locked Loops—Analogs Do All the Work of Digitals, and Much More." *EDN*, September 15, 1977, pp. 117–119.
19. Nemec, J. "Build a High-Frequency Synthesizer." *Electronic Design*, February 15, 1977, pp. 120–122.
20. Noll, E. "Circuits and Techniques." *Ham Radio*, September, 1971, pp. 54–59.
21. ——— "Circuits and Techniques." *Ham Radio*, October, 1971, pp. 58–61.
22. ——— "Circuits and Techniques." *Ham Radio*, December, 1971, pp. 70–71.
23. Pohlman, D. T. "Timer/Counter Chip Synthesizes Frequencies, and It Needs Only a Few Extra Parts." *Electronic Design*, June 21, 1974, p. 114.
24. Pulice, J. "Direct Output Two-Meter Synthesizer." *Ham Radio*, August, 1977, pp. 10–21.
25. Rasmussen, D. D. "A Tuning Control for Digital Frequency Synthesizers." *QST*, June, 1974, pp. 29–32.
26. Robbins, K. W. "Tunable Six- and Ten-Meter Phase Locked Loop." *Ham Radio*, January, 1973, pp. 40–44.
27. ——— "Six-Meter Frequency Synthesizer." *Ham Radio*, March, 1974, pp. 26–33.

28. Sharpe, C. A. "Speed Up PPLs." *Electronic Design*, November 22, 1977, pp. 124–127.
29. Stein, R. S. "Frequency Synthesizer for the Collins 75S Receiver." *Ham Radio*, December, 1975, pp. 8–27.
30. Stevens, D. H. "A 4000-Channel Two-Meter Synthesizer." *QST*, September, 1972, pp. 17–25.
31. Stinnette, N. "Phase-Locked Loop RTTY Terminal Unit." *Ham Radio*, February, 1975, pp. 36–37.
32. Tuhro, R. "Interface Ties Micro to Standard Cassette Recorders." *EDN*, October 20, 1977, p. 110.
33. Zwicker, R. M. "Phase-Locked-Loop Circuit Multiplies Frequencies by 2 to 256." *Electronic Design*, May 24, 1976, p. 94.

Index

A

Adder/decoder, 99

B

Basic
 principle of phase-locked loop, 7-9
 synthesizer, 79
 transfer system, 151-154
Basics of a vco, 53-54
Breadboarding, 13, 14

C

Capacitors, 21
Capture range, 71
CMOS
 fixed counters, 92-94
 integrated circuits, 21
 oscillators, 87-89
 phase-locked loop, 4046, 133-127
 programmable counters, 97-100
Cutoff frequency, 66

D

Damped natural frequency, 70
Debounced switch, 17-18
Divide-by-N counters, 90
 4018, 93
Divide-by-12 counter, 7492, 91
Double-balanced mixer, 123
Duty cycle, 28

E

Edge-triggered phase detector, 29-32
Equipment, 14-16
Error voltage, 26
Exclusive-OR phase detector, 27-29

F

560B phase-locked loop, 123-125
561B phase-locked loop, 125-126
562 phase-locked loop, 126
565 phase-locked loop, 127-129
567 phase-locked loop/tone decoder, 129-131

Fixed resistors, 21
Format for the experiments, 12
4017 and MM4617 decade counters, 92
4018 counter, 97
4018 presettable divide-by-N counter, 93
4046 CMOS phase-locked loop, 133-137
4522 counter, 99-100
40192 counter, 97
Frequency
 counter, 14
 reference circuits, 86-89
Fsk demodulator, 127-128
Function
 generator, 16
 of loop filter, 64-65

H

HCTRO320 digital frequency synthesizer, 98
Heterodyne-down conversion, 80
History of phase-locked loop, 9-10
Hold-in range, 70

I

Input/output circuits, 16-20
Integrated circuits
 CMOS, 21
 linear, 21
 TTL, 21

L

Lag-lead circuit, 66-67
LED monitors, 16
Linear integrated circuits, 21
Lock and capture, 70-73
Logic switch, 17
Loop filter synthesizer, 84-86
Low-noise detector, 134
Low-pass filter circuits, 65-67
LR-2 logic switch Outboard, 243
LR-6 lamp monitor Outboard, 245

253

LR-4 seven-segment LED display Outboard, 244
LR-7 dual pulser Outboard, 245
LR-25 TTL breadboarding Outboard, 246
LR-30 CMOS breadboarding Outboard, 246
LR-31 function generator Outboard, 246
LR-33 quartz crystal Outboard, 246

M

MC1648 voltage-controlled oscilator, 56-57
MC4016 counter, 96
MC4024 voltage-controlled multivibrator, 55-56 ·
MC4044 phase detector, 32-35

O

Omega, 54
Oscilloscope, 14
Overshoot, 67

P

Parallel varactor tuning, 58
Parameters for choosing varactor, 58
Phase, 24-26
   comparator, 26
   detector, 26-27, 134
      conversion gain, 26, 29
      experiments, 35-52
   difference, 25
   /frequency detector, 99
   -locked loop
      basic principle, 7-9
   /tone decoder, 567, 129-131
Pocket calculator, 14
Potentiometers, 21
Practical synthesizer, 79-84
Prescaling, 82
Programmable divider, 99
Programming switches, 100-103

R

Radian, 26
R-S flip-flop, 29
Rules for setting up experiments, 11-12

S

SCA decoder circuit, 129
Schematic symbol for frequency reference, 20
Series varactor tuning, 58
Settling time, 68
Set-reset, 29
7-segment LED display, 18
7490 decode counter, 90
74C90 decode counter, 94
7492 divide-by-12 counter, 91
74192 counter, 94-96
74C192 counter, 97
Solid-state devices, 22
Stable frequency reference, 18-20
Symbols, 249

T

Three states of phase-locked loop, 8
Tools, 13-14
Touch-Tone® decoder, 131-133
Tracking range, 70
Transient response, 67-70
TTL
   fixed counters, 90-92
   integrated circuits, 21
   oscillators, 87
   programmable counters, 94-97

U

Unit-decade-cascadable, 96

V

Varactor, 57-60
   diode, 57
   tuning
      parallel, 58
      series, 58
Vco
   circuits, 54-57
   conversion gain, 54
Vom, vtvm, digital voltmeter, 14-15

W

Wideband detector, 134
Wire, 14